Karstified Coasts of the South Andaman Islands

The lighthouse is located on the extended edge of the promontory rock body into the Andaman Sea floor at Havelock Island to guide the ships on the sea

Anurupa Paul · Jatisankar Bandyopadhyay · Ashis Kumar Paul

Karstified Coasts of the South Andaman Islands

Forms and Processes

Springer

Anurupa Paul
Department of Remote Sensing and GIS
Vidyasagar University
Midnapore, West Bengal, India

Jatisankar Bandyopadhyay
Department of Remote Sensing and GIS
Vidyasagar University
Midnapore, West Bengal, India

Ashis Kumar Paul
Department of Geography and Environment Management
Vidyasagar University
Midnapore, West Bengal, India

ISBN 978-3-032-10298-0 ISBN 978-3-032-10299-7 (eBook)
https://doi.org/10.1007/978-3-032-10299-7

This Springer imprint is published by the registered company Springer Nature Switzerland AG
The registered company address is: Gewerbestrasse 11, 6330 Cham, Switzerland

Preface

The volume entitled *Karstified Coasts of the South Andaman Islands: Forms and Processes* exhibits the first authored book on coastal karst features and their forms and processes reported from the Archipelago of Andaman and Nicobar group fringed with the Andaman Sea and the Bay of Bengal of the Northern Indian Ocean. Ideas on the speleogenesis were discussed among the authors during the field works (2015–2021) for a volume which will depict the regional reef limestone geomorphology in the tropical oceanic islands of the sub-continent. During the formulation of the scheme of the proposal for the volume, applications of coastal geomorphology, geospatial sciences, environmental science, physical oceanography, geological science and coastal ecology were considered to explore the forms and processes of karstified coasts in the isolated land masses. The reviewers of the *International Speleogenesis Centre (Karst Research Institute of Europe)* recommended the book proposal with strong and positive comments for the upcoming volume, in which the forms and processes of coastal karsts and island karsts will be reported from the region of the northern Indian Ocean.

The volume contains 5 splendid chapters on South Andaman Island, coastal geomorphology and habitats, karstified coasts and coastal hydraulics, reef islands and carbonate sediments, antecedent karst morphology and future responses to the changing environment. It is also noted that forms and processes of the karstified coasts are well represented by descriptions and analysis of a total of 34 issues and factors, containing South Andaman Island systems of the tropical oceanic environment. The antecedent karst morphology of the reef island platforms is explored and identified with the geological signatures and geomorphological features in the field associated with the application of geospatial techniques, spatial modelling techniques, and other existing geological records. Though there are limitations of research in the emerald islands but it will open the opportunities for further research on the karstified coasts of the archipelago. The book will provide a ready reference and datasets for the intended audiences of academics, professionals and classroom teaching on earth system sciences, coastal geomorphology, speleogenesis, physical oceanography, environmental science, geospatial science, geological studies and tourism travel management. However, the volume contains the literature of 234 pages, 34 tables, 182 figures and other illustrations to support the forms and processes of coastal karsts in the Andaman Islands for the above-mentioned audiences.

The researched volume will reflect the long experiences of the workers in the field of coastal studies and joint efforts of the authors from disciplines like Physical Geography, Geological Science, Environmental Science and Geospatial Science to furnish the work as per the proposal partially based on Ph.D. research and extended research on coastal karstification processes and resultant features of the Andaman Islands. We hope our expectations will be fulfilled by the attractive and splendid volume, as ready documents and databases for the interested students, researchers, professionals and travellers from different institutes with various motives, as well as from outside the country.

Midnapore, India

Dr. Anurupa Paul
Dr. Jatisankar Bandyopadhyay
Dr. Ashis Kumar Paul

Acknowledgements

The ideas for the book volume entitled *Karstified Coasts of the South Andaman Islands: Forms and Processes* were made in 2017 during monitoring shoreline geomorphology of the islands in the South Andaman district with my co-authors and other researchers. Distinctive karst features of the coastal fronts and island interiors suddenly attracted us to think and formulate a proposal for materializing the ideas of such daring work from the emerald islands. We congratulate Springer Nature, Switzerland AG, for accepting our book proposal during February 2024, and also convey our gratitude to the reviewers from Karst Research Institute, Europe (Slovenia, ZRC-SAZU) for their strong and positive comments about the upcoming volume. We also extend our gratitude to Zachary Romano, Senior Publishing Editor of *Earth Sciences*, *Geography*, and *Environment* at Springer, New York, and Climent Wilson, Production Editor, Springer Nature, for organizing the publication processes of the current volume.

We appreciate the assistance of Dr. Joydeb Sardar, Head of Computer, Faculty member at 'ANURUPA'S ACADEMY', Solapur, Maharashtra, India, in all stages of the work. He is especially acknowledged for his cooperation and assistance throughout the research. We would also extend our thanks as a token of appreciation for the cooperation and help provided by Mrs. Mousumi Paul and Advocate Surya Umbarje in the entire work, as well as their field assistance to fulfil our dream project.

We extend our gratitude to Mr. Kshitish Karak, Tour Agent for the Andaman and Nicobar group of islands, for his help in conducting field work in different destinations of the emerald islands from 2012–2021. Dr. Sujit Dasgupta, Deputy Director, GSI, Kolkata, for his extended help and cooperation in the field work during 2018 in the South Andaman Islands, and for which he is also highly acknowledged for the same. We would also like to extend our gratitude to the HODs of the Remote Sensing and GIS department and the Geography and Environment Management department of Vidyasagar University, Midnapore, West Bengal, India, for providing permission to use the digital laboratory and cartographic equipment during the stages of field work.

Dr. Anurupa Paul
Dr. Jatisankar Bandyopadhyay
Dr. Ashis Kumar Paul

Contents

About the Authors

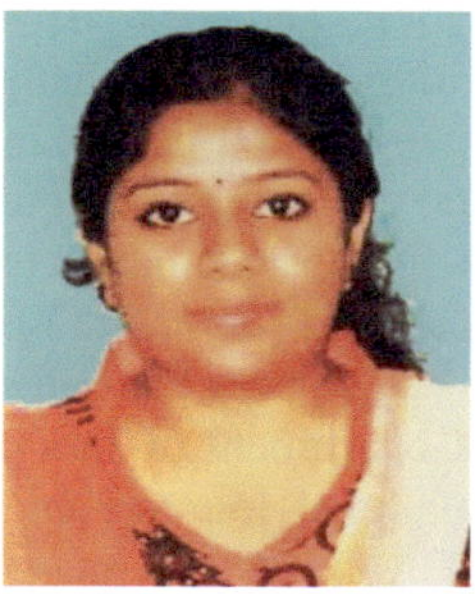

Dr. Anurupa Paul has completed her M.Sc. in Geography from Savitribai Phule Pune University, Maharashtra, and obtained her second M.Sc. degree in Environmental Science from Vidyasagar University, West Bengal. She has been awarded the Ph.D. degree from the Department of Remote Sensing and GIS of Vidyasagar University. She was awarded the title of Young Scientist in 2013 from Savitribai Phule Pune University, India. She has published 26 articles in reputed international and national journals and two book chapters in Springer Nature and Taylor & Francis. Anurupa has presented her research work in Kuala Lumpur, Malaysia (the 38th ACRS Conference), other international conferences (12) and 34 national conferences. She has authored the book as *Crisis on the Coast and Hinterland*. She has also participated in 36 workshops, international training programmes and 121 webinars on the application of remote sensing techniques in coastal research and disaster management. She has received seven best paper presentation awards in seminars. Her area of interest in research is mainly focused on Karst Morphology, Coastal Geomorphology, Limestone Geomorphology, Environmental Geoscience, Earth Science and the application of remote sensing and GIS. She has guided 18 M.Sc. dissertation in coastal geomorphology and environment with the application of geospatial techniques in the Department of Remote Sensing and GIS, Vidyasagar Universities. She is now currently doing her research activities in her own academy. She is also engaged with some environmental and social awareness campaigns in Pune-Konkan-Solapur Region, Maharashtra, India.

Dr. Jatisankar Bandyopadhyay has been working as Associate Professor and Head, Department of Remote Sensing and GIS, and Director, Centre for Environmental Studies, Vidyasagar University, West Bengal. He has completed his M.Sc. in Applied Geology from University of Allahabad, and M.Phil. and Ph.D. from School of Environmental Science, Jawaharal Nehru University, New Delhi. His broad area of research includes Geology, Earth science, Disaster Management, Environmental Impact Assessment (EIA), Environmental Studies and Application of Remote Sensing and GIS Modelling. He has supervised 10 (awarded), 5 (ongoing) Ph.D. scholars, M.Phil. students (awarded 2) and more than 160 M.Sc. students for dissertation work. He has published more than 116 articles and book chapters in the reputed national and international journals and edited book volumes. He has presented papers in more than 44 national and international conferences. He has participated in 40 workshops and 15 webinars. He has received 3 best paper awards. He has completed 6 projects and is engaged with 1 ongoing project as Principal Investigator with INCOIS Government of India. He is a reviewer of many international journals and a member of INCOIS, SAC, INCA, ISRS and many other reputed organizations. He promotes a stimulating learning and development environment with his vast teaching and research experience of over 28 years, verbal communication and excellent administrative and leadership skills.

Dr. Ashis Kumar Paul is a retired professor of Physical Geography, and he was teaching in the Department of Geography and Environmental Management of Vidyasagar University, Midnapore, West Bengal, India. He has visited the Archipelago of Andaman and Nicobar groups from 2003 to 2018 once a year for field work with research scholars and scientists of different institutes for the studies on physical and environmental aspects of the island shores. Geomorphological features and the process-forms interactions were monitored by him before and after the great Andaman earthquake and tsunami event (2004) in the island systems. He has also guided three Ph.D. research scholars on the geomorphology and environmental issues of the Andaman group of islands. Professor Paul has contributed to the monitored field assessment survey report of karstification processes and other marine hazards for the present work. He has also played a major role in writing manuscripts for the current authored book. He has authored the books *Coastal Geomorphology and Environment*, *Tsunami: An assessment of Disasters*, *Tectonics and Landforms*, *Physical Environment* and edited the volume *Crisis on the Coast and Hinterland*. Professor Paul has over 180 research papers, book chapters and articles in international, national journals, edited volumes and monographs to his credit. He has supervised 31 Ph.D. research scholars on coastal geomorphology, coastal management, coastal sediments, geoarchaeology, remote sensing and GIS, and coastal environment. He was associated with reputed national and regional projects as co-investigator and Principal Investigator (PI) with other scientists (1994–2015). Professor Ashis Kumar Paul was the president of the Indian Institute of Geomorphologists (IGI) in 2020 and was awarded membership as an authorized CERF Society Member of JCR, USA in 2017. He was also teaching in Remote Sensing and GIS (2007–2018) and Earth System Sciences (2019–2021) at Vidyasagar University. He is also reputed as a field geomorphologist and field ecologist. Professor Paul participated in several geoarchaeological exploration programmes and collaborated with archaeologists, geologists and earth scientists of different institutes for neolithic and palaeolithic studies in the coasts and coastal hinterlands of West Bengal and Odisha, India.

Abbreviations

ASF	Alaska Satellite Facility
BHB	Bay Head Beach
BHD	Bay Head Delta
BMB	Bay Mouth Bar
BSB	Bay Side Beach
$CaCO_3$	Calcium Carbonate
CB	Cuspate Bar
CD	Cuspate Delta
CGWB	Central Ground Water Board
CH	Cuspate Headland
CS	Cuspate Spit
CV	Cove
DEM	Digital Elevation Model
DSAS	Digital Shoreline Analysis System
ERDAS	Earth Resources Data Analysis System
ESA	European Space Agency
ETM+	Enhanced Thematic Mapper Plus
FCD	Forest Canopy Density
FCM	Forest Canopy Model
FLAASH	Fast Line of Sight Atmospheric Analysis of Spectral Hypercubes
FQAI	Forest Quality Assessment Index
GEBCO	The General Bathymetric Chart of the Oceans
GIS	Geographic Information System
GSI	Geological Survey of India
HB	Headland Beach
HTL	High Tide Line
I	Inlet
IMD	India Meteorological Department
INCOIS	Indian National Centre for Ocean Information Services
JNRM	Jawaharlal Nehru Rajkeeya Mahavidyalaya
L	Lagoon

LAI	Leaf Area Index
LTL	Low Tide Line
NATMO	National Atlas and Thematic Mapping Organisation
NBSS & LUP	National Bureau of Soil Survey and Land Use Planning
NE	North-East
NNW	North and North-West
NOAA	National Oceanic and Atmospheric Administration
NS	North-South
NW	North-West
OLI	Operational Land Imager
P	Promontories
RMSE	Root Mean Square Error
RS	Richter Scale
SAC	Space Application Centre
SAM	Spectral Angel Mapper
SE	South-East
SOI	Survey of India
SRTM	Shuttle Radar Topography Mission
SSE	South and South-East
SST	Sea Surface Temperature
SW	South-West
TIN	Triangulated Irregular Network
TIRS	Thermal Infrared Sensor
TM	Thematic Mapper
UNESCO	United Nations Educational, Scientific and Cultural Organization
USGS	United States Geological Survey
YBP	Years Before Present

List of Figures

List of Tables

Chapter 1
South Andaman Islands

1.1 Physical Setting of the Oceanic Islands (Northern Indian Ocean)

The Andaman group of islands is part of the accretionary prism of Island Arc setup along the line of subduction plate boundary between Indian Ocean Plate and Burma Plate (Regional Plate of Eurasian Plate). The island arc is extended from north to south as 700 km long ridges within 10° N to 13° N latitude with number of segments separated by intervened channel straits or channel creeks in the region of Bay of Bengal and Andaman Sea. Cross-sectional forms of the Andaman group of islands represent ridge and valley topography from west to east (92° E to 93° E longitude). Large volume of deep-sea sediments and oozes were uplifted, tilted, folded, faulted and sliced in the accretionary prism under the plate tectonic stress field in the geological time between Cretaceous to Late Quaternary period. Ophiolites and other group of igneous rocks are lying at the basement of sedimentary volumes in Andaman group of islands. There are three significant North–South oriented fault lines, viz. Jarwa Fault, Diligent Fault and Eastern Margin Fault (Fig. 1.1) in the region which separated the island mass into parallel ridges and valleys. Havelock Island and Neil Island are located on the east of Eastern Margin Fault in Andaman Sea and Little Andaman Island extended on the west of diligent fault. Little Andaman is bounded by the Bay of Bengal on the west and Andaman Sea on the east, and it is separated from the main group of islands by Duncan Passage and Ten-degree channel.

The study areas of three selected islands belong to South Andaman district and geologically exhibit a separate status with Archipelago Formation in comparison with other islands of the main group. Such formation is mostly dominated by carbonate rocks and reef rocks. Topographically, these islands represent denudational ridges, parallel or linear hill ridges, intermontane valley flats, reef terraces, sea cliffs, island fringe shore platforms, ria type shores, headlands, coves, promontories, beaches, etc., land units. Islands are elevated up to 100 m from the mean sea level, and densely covered by tropical monsoon forests, semi-evergreen forests, evergreen

A. Paul et al., *Karstified Coasts of the South Andaman Islands*,
https://doi.org/10.1007/978-3-032-10299-7_1

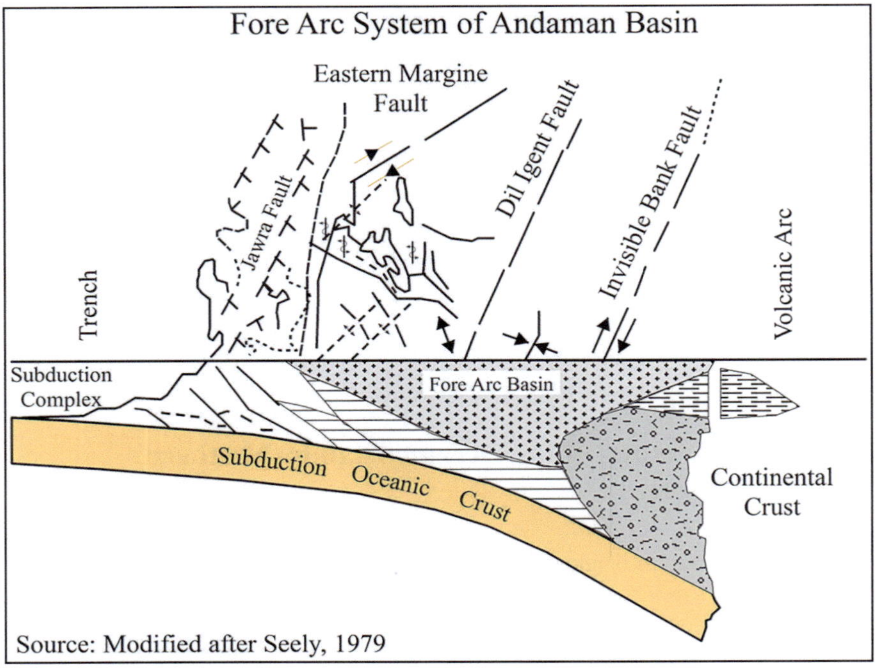

Fig. 1.1 Active tectonic zones of the Andaman Basin in North Indian Ocean

forests, littoral forests and mangroves along the estuary fringes and embayments. Active drainage channels are extended outwards of the central highlands of Little Andaman and they discharge their runoff directly into the sea. There are six major drainage channels in Havelock Island which exhibit sluggish in nature particularly along the swampy terrain in the valley flats and basinal surface. Various types of springs are found along the coastal hills, island inter hills and also on the sea cliffs during rainy months (April –October) in the island. However, there is no significant drainage channel in the porous rocks of Neil Island but represent as many as 19 seepage flow channels (Fig. 1.2) along the shore fringes of the region.

All three islands are fringed with coral reefs at the depth ranging from 10 to 25 m (Havelock and Neil) and 20–50 m (Little Andaman) bathyal zones of Andaman Sea and Bay of Bengal. The windward sides of Neil and Havelock Island accumulated thick deposit of reef materials with coral rubbles, gravels and boulders. Phytokarstic features and beach rocks as well as sediment cays occupy the shore platforms and shore cliffs of the island fringes, and they provide opportunities for exploration of bio-karstic processes in the evolution of these carbonate islands under tropical coasts. Coastal processes like tides, swell waves, cyclone surges and tsunamis modify the morphology of shore platforms by supply and transport of sediments, by sub-aerial weathering and wave abrasion, and also by cliff recession activities in the region. The

Fig. 1.2 Location of South Andaman Group of Islands

Pleistocene-Holocene Sea level fluctuations and seismo-tectonic activities played significant role in the evolution of island systems.

1.2 Physical Conditions of the Study Areas

The present section deals with the physical makeup of islands in the region of Andaman Sea and Bay of Bengal in consideration with the characters of climate, geology, hydrology, topography and soil types.

1.2.1 Climate (Environmental Variations)

Climate represents the average weather condition of 35 years period of a region. The elements of climate are discussed (55 years of data are considered) in the following section to represent the environmental variations of the islands.

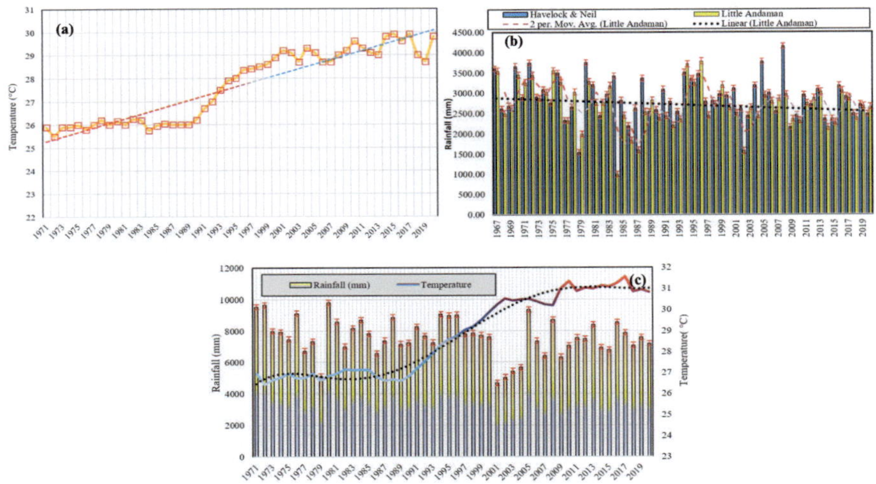

Fig. 1.3 Multiple curves showing the slow rate of declining trend with short-term events of rainfall and increasing trend of temperatures (**a**), (**b**), and (**c**) over a 50-year period at the Port Blair station (South Andaman) based on the IMD data source

1.2.1.1 Temperature

The temperature, as a climatic parameter plays a significant role in the mechanical disintegration of rock bodies and also acts as an agent in biochemical degradation of materials along the jointed rocks. Average temperature data of 38 years, i.e. from 1949 to 1986 represents seasonal fluctuation and differences between minimum and maximum temperatures of South Andaman Islands. The minimum temperatures range from 21 °C (February) to 23.9 °C (May) in the period of 12 months' time. However, the maximum temperatures of the similar time period exhibit the range between 29 and 32.6 °C in South Andaman (Fig. 1.3a). The differences between minimum and maximum temperatures also range from 7 to 8.5 °C in the region within a year. The sea surface temperatures (SST) of northern Bay of Bengal and Andaman Sea achieve 29.5–31 °C during April–May (2018–2020). However, the annual average SST of the region achieved around 29 °C during the previous decade (2010–2020), which has rendered the area prone to development of cyclones.

1.2.1.2 Rainfall

The average annual rainfall of the region is estimated as 3086.8 mm over 38 years period between 1949 and 1986. Maximum rainfall occurs during the months between May and December, and number of rainy days varies from 13 to 21 in the same period. The data also reveals that maximum amount of rainfall was registered in the months of June, July, August and September (> 460 mm). June was the rainiest month

(498 mm) in South Andaman Islands, however, the events of storm rainfall (2016, 2018) registered 216 mm and 166 mm in 24–30 h period in the previous decade during the month of January–February (Fig. 1.3a, b).

Rainfall plays the most important role in the distribution of natural forest types, and acts as a major agent of chemical weathering over the carbonated rocks in the island areas. Drainage systems of Little Andaman and Havelock Island are fed by rainfall. The rate of infiltration and surface runoff are usually guided by amount of rainfall, forest cover areas and topography. The self-sustaining capacity of 'Rain Forest Hydrological Cycle' usually operates the mechanism to produce local rainfall in the pre-monsoon summer months (February–April) in the island region. Base flow of the drainage channels is also influenced by occurrences of rainfall during the period between May and December months in the islands.

1.2.1.3 Significance of Windward and Leeward Sides of Island

The island fringe shore platforms and fringing coral reefs are largely influenced by windward sides and location of leeward sides. Windward sides receive high wind speed and wind-driven waves which favours better growth of coral reefs and strong wave erosion. Finer sediments are accumulated on the leeward sides of the island (Neil), and larger grains, i.e. gravels, rubbles, cobles and boulders drifted and transported across the shore platforms of windward sides.

Large-scale marine inundations with wind-driven waves have produced wave abrasion on the shore platforms and active cliff recession by advanced wave breakers on the windward sides (Fig. 1.4). The islands of Little Andaman, Havelock and Neil, however, are facing windward sides seasonally towards southeast (April–October) and also towards northeast (November-March) due to tropical monsoon wind movements through bi-directional system in the northern Indian Ocean. Shore platform geomorphology usually becomes dynamic with sediment movements and strong incoming wave energies on the windward sides of the islands.

1.2.1.4 Wind Speed and Wind Direction

Wind speed ranges from 12 to 20 km per hour during the monsoon months (May–September) in a normal condition, but in other months (October–April) it varies from 6 to 9 km per hour in the oceanic region. The direction of wind is mostly guided by southwest monsoon and northeast monsoon winds. However, in the period of cyclone phases wind approaches the island fronts from all sides with significant speed (> 100 km/hrs.). Wind speed and wind direction plays the most important role in building the local waves and distant waves (fetch waves), and also increase the wave amplitudes with high velocities. The direction of wave approaches is also guided by wind direction (Wind Rose Diagram), and mostly dominated by north, northeast, and east winds during the months of March and April in the islands. However, the wind direction and wind speed changes with seasonal shift, and they

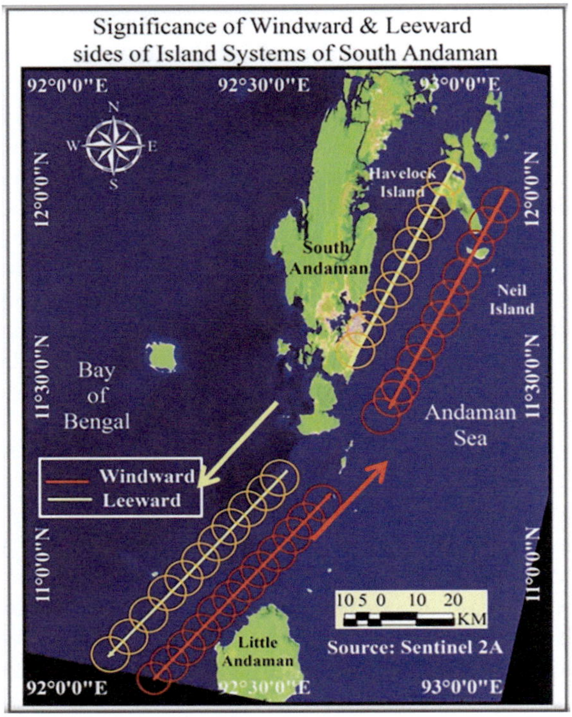

Fig. 1.4 Windward and leeward sides of South Andaman Islands

approach towards the island shores from west (64%), southwest (25%), south (4%) and northwest (4%) during the month of September with high velocities (Fig. 1.5).

1.2.1.5 Relative Humidity

The average annual humidity in the atmosphere for South Andaman Islands is estimated as 80.25% on the basis of available data from IMD. Relative humidity measured at 8.30 a.m. and 5.30 p.m. also vary in the islands. The monthly average humidity is showing relatively higher at 5.30 p.m. than the morning time (8.30 a.m.) as rapid heat transfer takes place in the day time through evaporation and evapotranspiration in the forested islands with continuous supply of moistures from the sea. Monthly humidity is relatively higher (>86%) during July, August, September, and October, and showing the lower rate of relative humidity (70–76%) in the months of December, January, February, March and April in the oceanic islands. The relative humidity however, rapidly increases (>90%) during the phases of cyclones to develop the storm rainfall around the areas of islands. The 'Rainforest Hydrological Cycle' is run by ready supply of moistures through relative humidity in the air (Fig. 1.6).

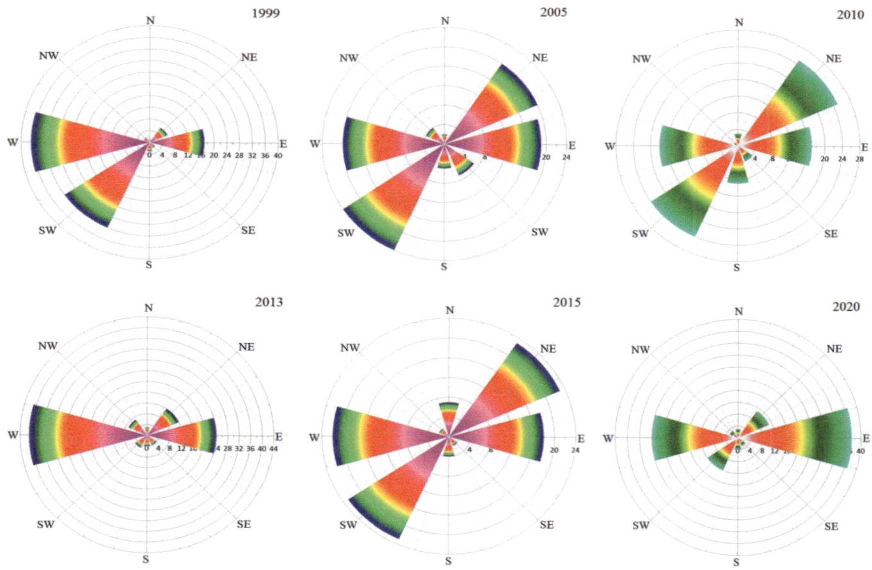

Fig. 1.5 Wind rose diagrams represent the dominant wind directions of Port Blair station (1999–2020)

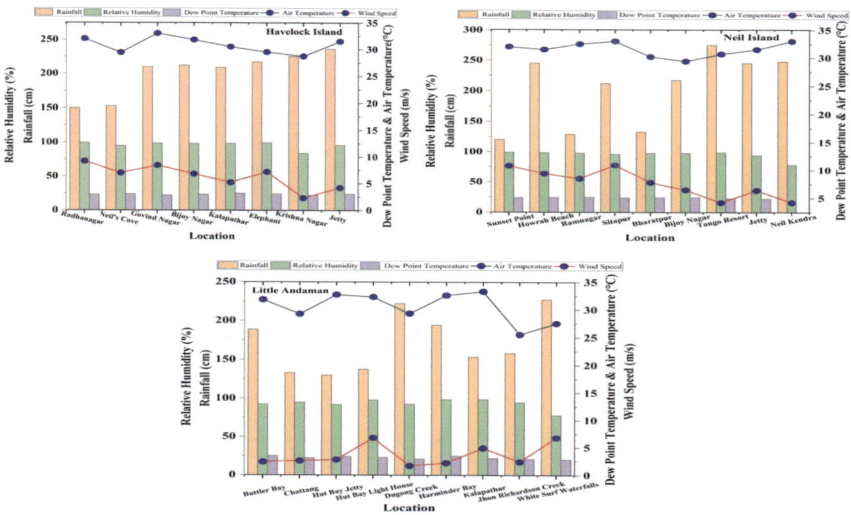

Fig. 1.6 Histograms and line graphs of environmental variables (rainfall, temperature, relative humidity, dew point temperature, and wind speed) showing the significant change in Neil and Little Andaman based on field data

1.2.1.6 Dew Point Temperature

The dew point is a stage of temperature for which the air usually becomes cooled to become saturated with the supply of water vapour. If the dew point becomes cooled further, the supply of airborne water vapour will condense to develop as liquid water or dew. If the air cools to its dew point with the contact of forest canopy surface that is relatively colder than the air, the water will condense on the top of the dense forest. A higher dew point indicates the presence of more moisture in the air (Fig. 1.6).

1.2.1.7 Spectral Temperature

The spectrum of the sun's temperature and wave length are absorbed differently by the nature of objects on the earth (Fig. 1.7). The spectral temperature of the coastal features on the shore fringe area of different islands is monitored using the infrared thermometer to make a spectral library for different features in the oceanic islands. The debris of stony coral, the calcareous sandstones on the cliff wall and the dry siliceous sands and shelly sands recorded the maximum spectral temperatures on the sea shore. However, the mangrove plants (*Avicennia* sp.,*Rhizophora* sp.), *Ipomoea* sp.,*Pandanus* sp., grassland vegetation, mangrove stilt roots, living coral, sea water recorded lower values of spectral temperature in the sea face. The remaining features like coral debris, wet sands, Phytokarst surface and beach rock showed moderate values of spectral temperature. Such monitored information of spectral temperatures is utilized for the validation of satellite image data.

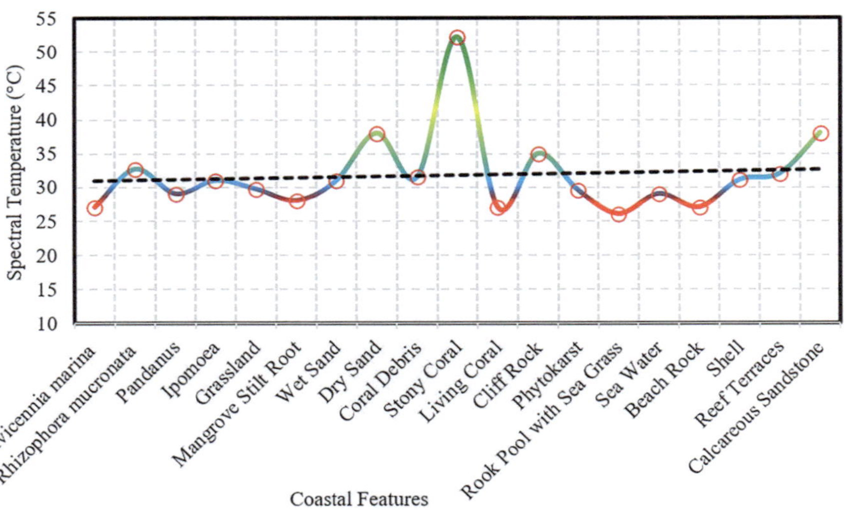

Fig. 1.7 Spectral temperature curve prepared based on monitored data in the study areas

1.2.1.8 Atmospheric Pressure

Atmospheric pressure is a unit area of air pressure exerted on the surface of the earth due to gravity pool of the atmospheric blanket on the globe. Atmospheric low pressure leads to cloudiness, wind, and precipitation, and high pressure system of the local atmosphere leads to usual fair and calm weather situation in an area. Atmospheric pressure gradient fluctuates with the changing temperature and elevation of the land (Fig. 1.6). The present study areas are located in the transitional zone between Equatorial Low Pressure belt and North East Trade Wind belt of Low Latitude region (10° N–12° N) over the Northern Indian Ocean. The atmospheric pressures become 1011 mb to 1007 mb during the month of April, and when tropical cyclones occur in the monsoon, the pressure decreases to 995 mb to 1000 mb (September).

1.2.2 Geology of the Study Area

Geologically, the islands belong to the 'archipelago formation' which constitutes various types of carbonate rocks and limestones of Holocene to Pleistocene age, Pliocene to Miocene age and Oligocene to Eocene age. The top surface is blanketed with alluvial and colluviums, mangrove swamp clay, raised beach materials, coralline sands and fragmented coral rags. The next underlying layer constitutes Melville limestone, round chalk, white clay stone, strait formation and calcareous sandstone. The bottom layer is composed of sandstone, siltstone, conglomerate grit and shale limestone in the island stratigraphy. The rocks are jointed, tilted, fractured, sheered, fissured and crashed under the stress field of tectonic boundary. Normal faulting in the rock beds is also recorded along the shore cliffs of Neil Island and Havelock Island (Fig. 1.8).

The soluble rocks and brittle rocks, the alternate hard rocks and soft rocks, and the nature and angle of bedding planes exhibit the erodibility rate of shore cliffs by wave erosion and mass wasting process. Beach rocks and reef rocks of the shore platforms are karstified with bio-encrustation, chemical weathering and active solution process in the tropical climate.

1.2.3 Soil Types

The soil types of these islands are categorized into *fine typic humitropept, loamy skeletal fragmental typic eutropept, loamy skeletal fragmental typic troporthent, fine loamy haplic hydraquent, and fine loamy aeric tropaquept (*Fig. 1.9*)*. They are exhibited by topographic diversity as medium hill region, low hill ridges, steep slopes of the hill ridges, mangrove swampy terrain, and along the valley flats of intermontane areas of island interiors. Among them the fine loamy aeric tropaquept is productive and cultivable in the region (Figs. 1.9 and 1.10).

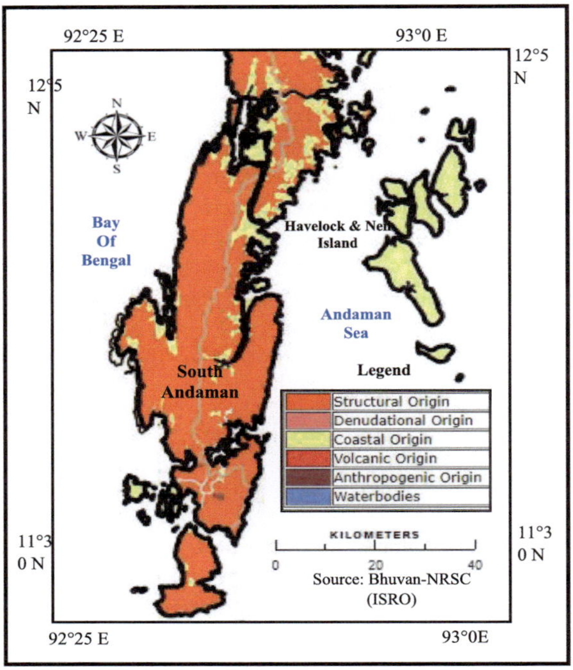

Fig. 1.8 General physiographic features and anthropogenic features of South Andaman

1.2.4 Drainage Characters

All three carbonated islands are drained by rain-fed channels and seepage outflow channels under tropical warm and humid climate. Large number of drainage channels with their tributaries flowing outwards from the central highland topography (> 100 m elevation) in Little Andaman Island. The long profiles of major drainage channels showing 3–4 knick points along their courses as they are crossing several terraces and abandoned cliffs within the island system. Waterfalls and rapids are well developed on the limestone cliffs along the courses of drainage channels. Among them, the 'white Surf' waterfall and 'Whisper' waterfall (Fig. 1.11) are important sites of geo-tourism destination in the island under dense forest cover surfaces. All the major drainage channels have produced 'Ria-type' estuaries along their coastal sections by subsidence and submergence process in the geological past. Mouths of the drainage channels are moderately influenced by high tides in the region. The ephemeral drainage channels become active with large volume of water flow during the monsoon months (June-October) and transport graveliferous along their beds in the region (Fig. 1.12).

The hilly tracts of Havelock Island (>90 m elevation) are also identified as the sources of several drainage channels in South Andaman district. There are four

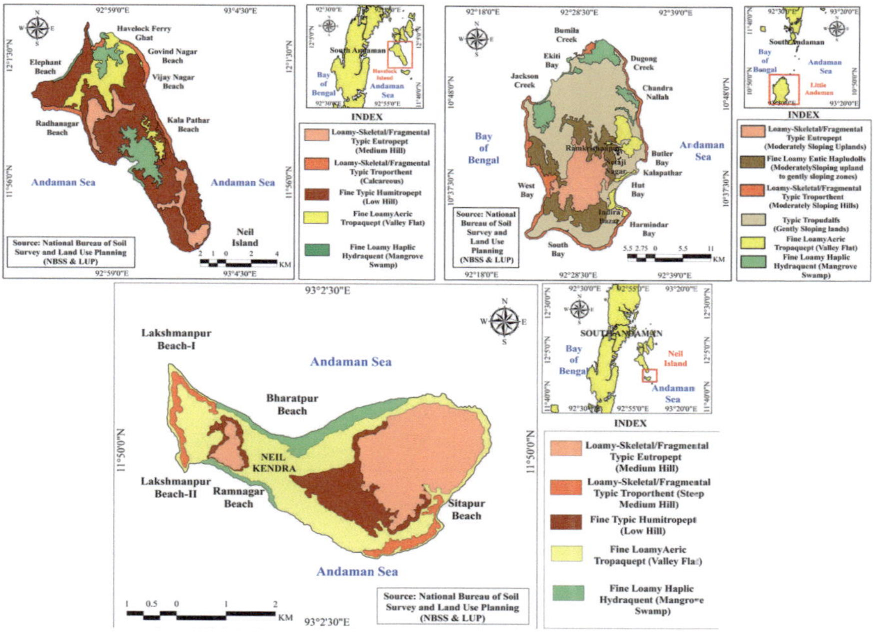

Fig. 1.9 Regional diversity of soil in Havelock Island, Little Andaman and Neil Island

major drainage channels in the interior parts of the island and about (10) ten smaller drainage channels (swift flowing) across the descending slopes of the coastal hills in the island. Among them, three major channels reached the south-central basin across the hilly course, basin margin flats with sluggish course and swampy course in the island (Fig. 1.12). The northern drainage channels with their tributaries flow towards the Andaman Sea across the intermontane valley flats and embayed coasts with meander courses and 'Ria-type' shores or estuary sections.

Neil Island has no significant surface drainage system but there are (19) nineteen number of seepage outflow channels across the shore platforms. The porous nature of carbonate rocks with coral fragments and other carbonate materials did not support the significant surface drainage channels due to infiltration and runoff processes in the region. However, two major surface depressions on the lower most reef terrace towards north and south (adjacent of Govind Nagar and Neil Kendra) accumulated seepage waters to support the mangrove wetlands in the region (Fig. 1.12).

1.2.5 Hydrologic Characters

The pseudokarst features, carbonate platforms and drainage features are evolved through hydrological characters of seepage flow, infiltration rate, underground flow

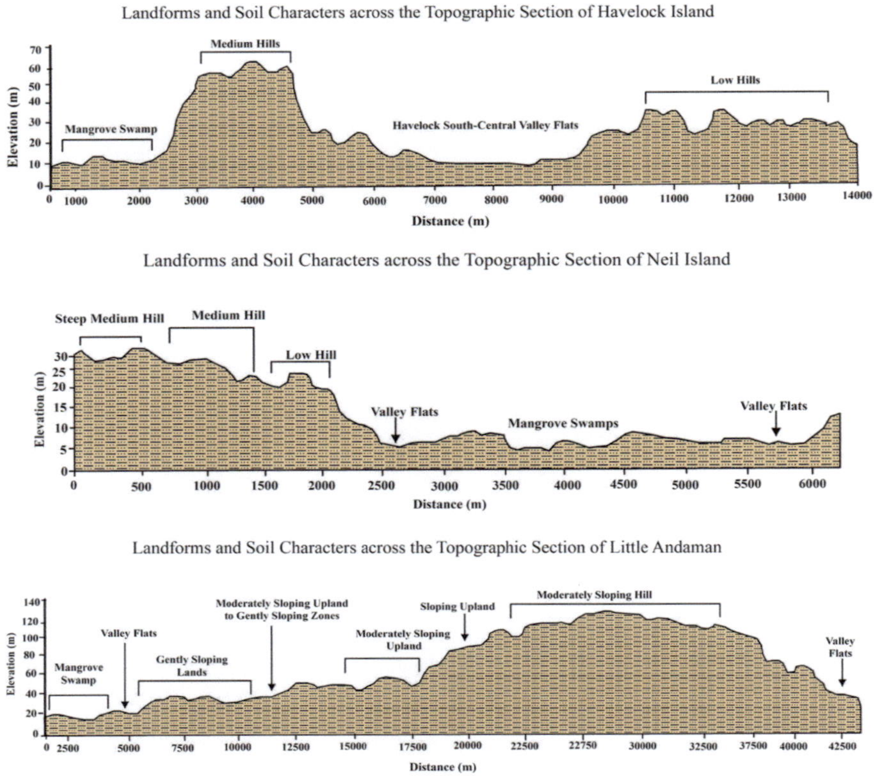

Fig. 1.10 Topographic forms are thematically modified and associated soil pedons of Havelock Island, Neil Island, and Little Andaman

of drainage waters in places and solution process in the islands. The fresh water at the land sea interface fluctuates with tidal pumping along the HTL, and seeps through the island fringes with degassing of carbonates for producing beach rocks in the upper part of shore platforms in the region (Fig. 1.13a, b).

Large amount of rain waters (> 3500 mm per year) in the carbonated islands developed karst drainage system in the subcutaneous zone and saturated zone store with moderate amount of surface overflow as well as deep loss to the regional ground water system. The limestone terrain of Little Andaman is densely forested over 98% areas of the island surface, and provided high infiltration rate with continuum of percolation in the region of island interiors. The seepage lines along the banks of drainage creeks (*Dugong creek, Bumila creek, Jackson creek and Benyabol creek*) and along the upper limit of high tide shoreline of Buttler Bay, Hut Bay, South Bay and Harmindar Bay are clearly visible during the pre-monsoon months (February–April) in Little Andaman. Direct discharge of fresh waters into the channel creeks and open sea environment has generated several karstic features along the estuary banks and

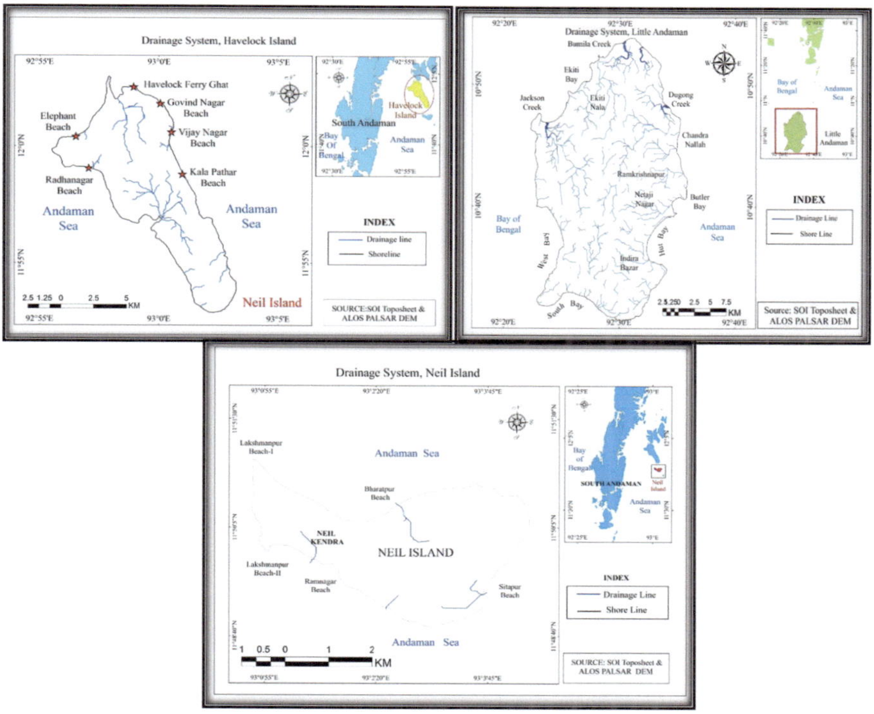

Fig. 1.11 Restricted drainage channels in Havelock Island, drainage map of Little Andaman, and Marginal drainage link channels of Neil Island

shore platform fringes. Surface overflow and drainage discharge during monsoon rains dominate over the surface detention, ground water recharge and evapotranspiration processes in the island systems (Paul, 1998). Seepage outflow channels and several springs emerged in the vadose and shallow phreatic zones exhibit the possibilities of active Karstification process in the islands (Fig. 1.13b).

In Havelock Island, seepage flow waters are emerged along the fringes of intermontane valley flats at the hill slopes to keep saturate the soil moisture of interior parts (Fig. 1.14). Wetlands with marshy tracts usually developed in the region for supply of such seeping waters through the sloping tracts, and used by the local people for cattle grazing and paddy cultivation activities. Seepage outflow channels on the shore platforms also provided opportunities for mangrove stands in the island. However, there are no active drainage channels over the sloping surface of carbonated Neil Island in South Andaman district.

Fig. 1.12 White Surf Waterfalls with limestone cliffs, weathered surface, and multiple knick points in Little Andaman (2015)

1.2.6 Contour Patterns

Contour plans are generated for all three islands with the help of ALOS PALSAR DEM to represent the topography. The patterns of contour represent central highland with descending surfaces of terraced landscapes in Little Andaman, linear hill ridges with intermontane valley flats and terraces in Havelock Island, and residual hill with raised beach terraces in Neil Island (Fig. 1.15).

Such diversity of topography is well represented in the cross profiles of island contours. Contour heights are varying from island to island as 135 m elevation is

Fig. 1.13 **a** John Richardson channel in South Bay of Little Andaman, **b** significant site of beach rocks in Lakshmanpur-II (Neil Island)

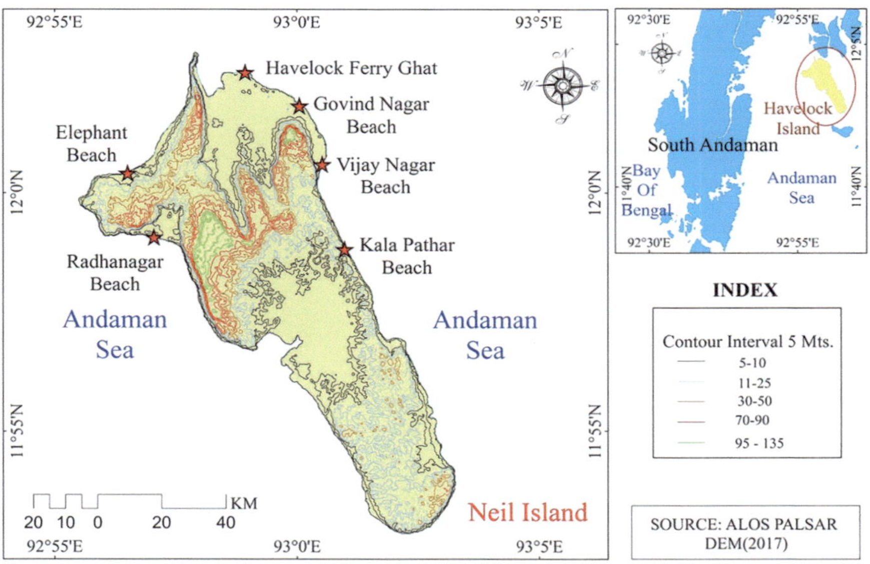

Fig. 1.14 Contour patterns of Havelock represent the nature of topographic forms

recorded in Havelock Island, 100 m elevation in Little Andaman and only 30 m elevation in Neil Island of South Andaman district. However, the lower most surface of Havelock and Neil Island represents 5 m contour, and Little Andaman on the other hand exhibits 10 m contour line along the coastline of the region. Thus, the events of large-scale marine inundations influenced more supply of external sediments input into the shore platforms of Neil and Havelock Island in comparison with Little Andaman (Figs. 1.14, 1.15 and 1.16).

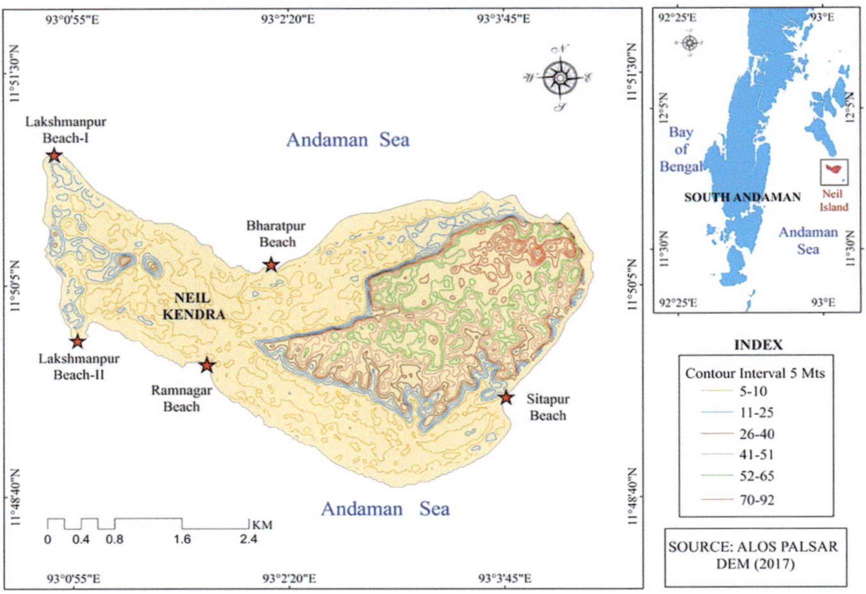

Fig. 1.15 The contour patterns of Neil Island represent the nature of topographic forms

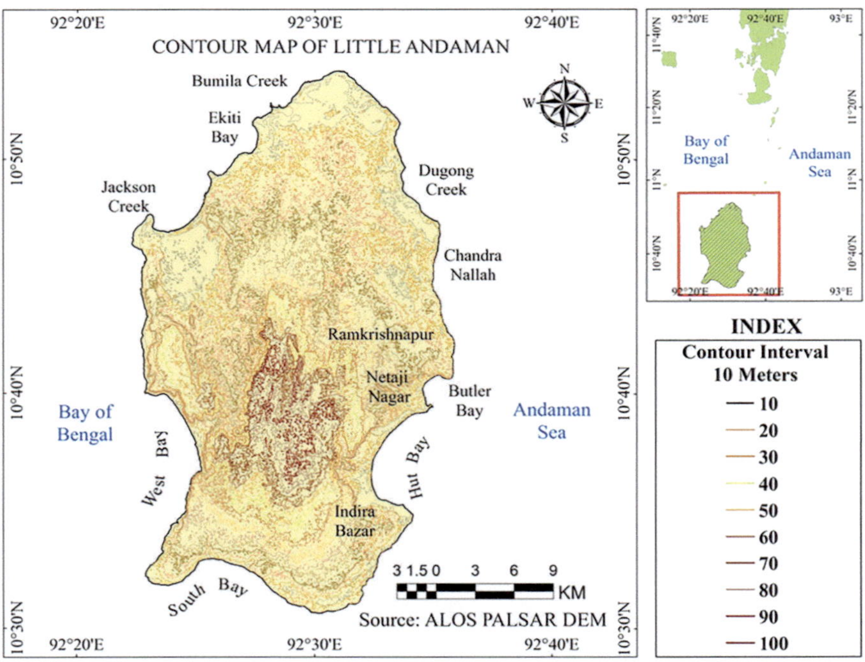

Fig. 1.16 Contour patterns of Little Andaman represent the nature of topographic forms

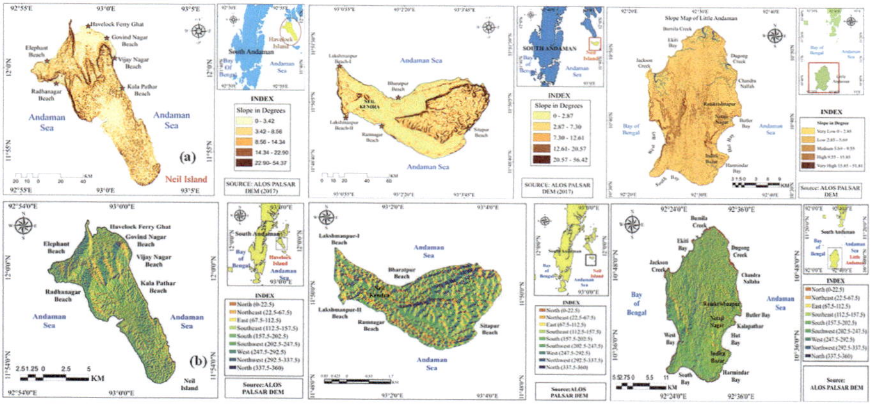

Fig. 1.17 **a** Havelock Island, Neil Island and Little Andaman Island exhibited high diversity of regional slopes and **b** contrasting slope aspects of Havelock, Neil Island and Little Andaman in South Andaman

1.2.7 Slope Characters

The slope map of Havelock Island exhibits 3.5°–54° degree surface slopes with significant break of slopes along the steeper hill fringes on the northern island interiors and western shores. The areas of valley flat and south central basinal surface show 3.5°–8.5° slopes in the island tract. The raised beach terraces of Neil Island exhibited only 3°–7° slopes, and flank margins of residual hill represented 21°–56° slopes respectively. The scarplands of Hut Bay and West bay represents the amount of very high slopes of 36°–52°, and the westward escarpment with undulated terrain exhibited high degree of slopes ranging from 10° to 16° in Little Andaman. The west flowing and east flowing drainage channels of the island exhibited with waterfalls and springs along their courses across the break of slopes in the terrain. However, the courses of drainage channels are sluggish on the low slope (3°–5.5°) terrain of the island (Fig. 1.17a).

1.2.8 Aspect of Terrain

The direction of slopes has guided the flow paths of channels. The outward slopes are directed towards east, west, north and south in Little Andaman. Overland flows on the island surface are clearly directed by outward slopes of the undulated terrain. Northward and southward slopes of Neil Island have guided the seepage outflow channels towards north, northeast and south in the region. The slopes of south central basinal topography are directed towards the inner parts of depressions in Havelock Island. The coastal hills on the west and north directed the terrain slopes towards west, east and northeast in the island. The intermontane valley flats and embayed shores of Havelock Island accumulated seepage outflow waters from the adjacent hill slopes. Sloping surfaces are directed towards north in the region of Havelock Island (Fig. 1.17b).

1.2.9 *Triangulated Irregular Network (TIN)*

Triangulated Irregular Network (TIN) of terrain conditions are product of elevation, slope and aspect. The three-dimensional views of the surface terrain are well represented in TIN map of Havelock Island. The diagram also exhibits four different geomorphic surfaces of the island with dissection of highland fringes. Similarly, the three geomorphic surfaces of Neil Island, and 5–6 major geomorphic surfaces of Little Andaman are clearly reflected in the TIN diagrams. The diagrams also highlight the presence of high diversity of geomorphic surfaces in Little Andaman in comparison with the other islands of the region (Fig. 1.18).

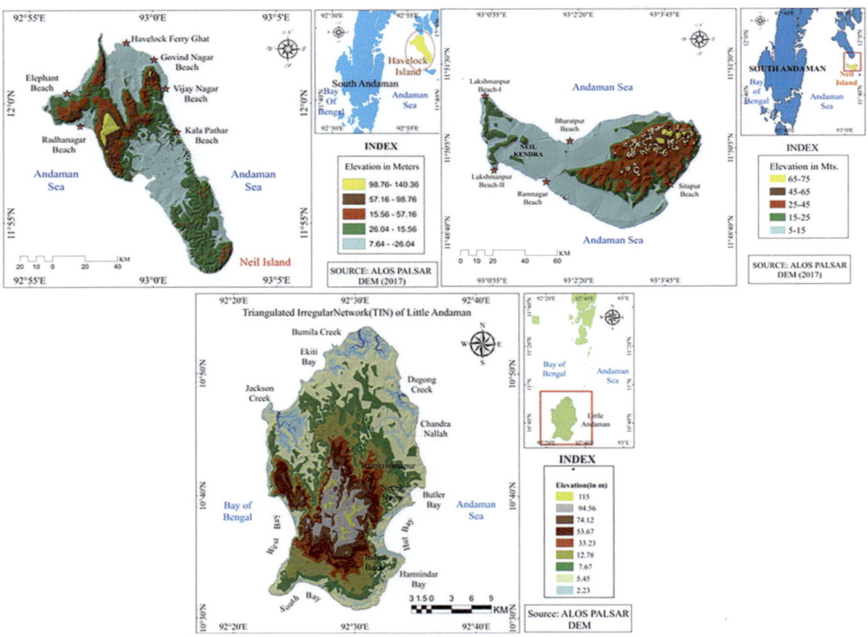

Fig. 1.18 3D effects of the physiographic features through TIN (Havelock, Neil Island and Little Andaman)

1.3 South Andaman District with Group of Islands

Location of an area represents the physical settings and climatic characters which can influence the rate of change in denudational processes of particular region. The present study areas are located in between 10° North latitude to 13° North latitude and between 92° East longitudes to 93° East longitude (Fig. 1.19). Tropical maritime environment of Andaman Sea and Bay of Bengal influences physical settings of isolated land masses by seasonal variation of winds, currents, swell waves and amount of precipitation. High rainfall, moisture supply, temperatures and evapotranspiration rate created an ideal condition for operating the active biological processes in the region. Island location on the other hand exhibits open marine environment from all sides of the shore fringes to distribute energy and materials along the shoreline configuration types.

Luxuriant growth of tropical rain forests, tall growth of mangrove stands with wide canopy covers (Fig. 1.20), decomposition of litters and extent of lichens and algal mats of the island platforms are favoured by tropical location and maritime influences. Locational significances of the present study areas are also reflected on

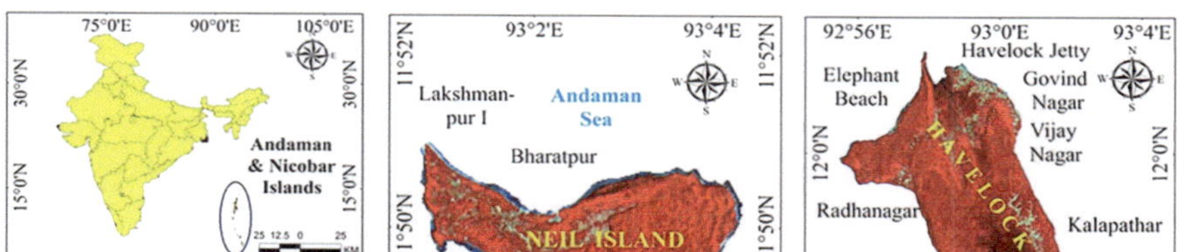

Fig. 1.19 South Andaman province constitutes three main islands, names as Neil Island, Havelock Island and Little Andaman

Fig. 1.20 Shore platform mangroves with stilt roots and lianas *(Rhizophora mucronata) on* the sea face of Neil Island showing the rainforest habit in the South Andaman district

the high rate of SST, events of cyclone breaks, and occurrences of storm rainfalls under tropical oceanic environment.

Stronger impacts of west coast swell waves and relatively moderate impacts of east coast swell waves are felt on the island shores for linear location of Andaman ridges from north to south in the low latitudes. Coral reefs are luxuriant and diversified on the west coast of Andaman Islands for above reason related with energy variation of winds and turbulence of sea waters.

Tropical moist climate of the region also favours condition for operating weathering process on the carbonate rocks of islands. Location of the island along the island arc setting is largely influenced by stress field of subduction-obduction boundary of the regional plate tectonics. Structural deformations of carbonate rocks are caused by such location of islands into the stress field.

The Andaman and Nicobar group of islands of the northern Indian Ocean are separated into six district units by administration purposes. South Andaman is the southernmost part of the Andaman Group of Islands and extended from Baratang Strait to the 10° degree channel in the Archipelago. Port Blair is the head quarter of the south Andaman district. The island systems are again grouped into three different sub-systems, such as i. Mainland island system; ii. Bay of Bengal fringed island system and iii. Andaman sea fringed island systems.

The mainland Archipelago constitutes a series of thrust belt ridges separated by linear valleys extended from north to south and fed by primary and secondary drainage systems. The ridges are elevated up to 400 m altitudes in places from the basement and deformed by the thrust belt tectonics. Linear valleys from north to south are occupied by 'Ria' type estuaries with broad intertidal and sub-tidal flats and filled by carbonate sediments. Ridges and valley are densely forested by evergreen trees, semi-evergreen trees, Littoral vegetations and swampy vegetations. The south Andaman at its central part is broken by embayments of North Bay and Flat Bay water bodies but the shore fringe areas represent extensive valley flats and terraces suitable for cultivations and settlements. Port Blair city, the capital of A& N group of islands is located at the shore fringe areas of central bay of the South Andaman district. Lithologically, the region constitutes basement ophiolites, Pleistocene limestones, Mithakari group of sediments and archipelago group of sediment and flanked by the Andaman Flysch sediments (Table 1.1).

The Bay of Bengal fringe shores and Andaman Sea fringe shores of the mainland archipelago are marked by embayed shores and broken shores with smaller bays, coves, headlands, promontories, and steep-sided valleys and cliffs. Coral reefs are fringed with the western and eastern of the mainland archipelago, but the wide shore platforms and reef terraces behind them reflect the quaternary history of the sea level oscillations and supply of carbonate sediments. Once the shore platforms and estuaries were occupied by mangrove swaps but demolished in many places after the great Andaman earthquake (2004) and associated tsunami events.

Islands of the Andaman Sea are facing towards east of the mainland Andaman ridge which depicts the younger archipelago group of rocks dominated by shell limestones, white clay stones, calcareous sand stones, beach calcarenites, reef rocks, chalky limestones, marls, etc., several karstic features are well developed in the shore platforms, shore cliff, and island platforms of the Andaman Sea. Topographically, they constitutes reef terraces, limestone ridges, intermontane valley flats, channels fringe marshes, limestone pavements, embayed shores with coves, bays, promontories and beaches. Island interior platforms are densely vegetated by semi-evergreen

Table 1.1 South Andaman Island systems of the archipelago

Mainland Island System (1)	Andaman sea fringe Island System (2)	Bay of Bengal fringe Island System (2)	Remarks (Topography)
Chidiya Tapu	Outram Island	Defence Islands	(1) Ridges, Valley flats, Ria shores, Hill slopes, drainage channels
Wandoor	Strait Island	Alexandra Island	
Vijayapuram	Henry-Lawrence Island	Tarmugli Island	
Viper Island	Sir William Peel Island	Red Skin Island	
Ross Island	Havelock Island	Maly Island	
Colinpur	Neil Island	Hob Day Island	(2) Reef terraces, Intermontane Valley flats, Hill slopes, Marshes
Bamboo Flat	Ross Island	Boat Island	
Wimberlygunj	Rutland Island	Jolly Buoy Island	
Shoal Bay	Cinque Island	North Sentinel Island	
Khurmadera	Passage Island	East Twin Island	
Jarawa Reserve	Sisters Island	Rutland Island west	(3) Reefs and reef terraces, Shore cliffs, Hill ridges
Nayagarh	Brothers Island	Cinque Island West	
Kanchangarh	Little Andaman	Sisters Island West	
Baratang		Brothers Island West	
		Little Andaman West	

trees, littoral vegetations and swampy forests. However, the shore fringes are modified and reworked by active marine processes in the tropical coasts. Coral reefs are fringed with island shores and they supply carbonate sediments into the shore platforms by wind waves, storms and tsunami events. Surface deformations by active thrust belt tectonics and seismo-tectonics represented jointed limestones, fissures, fractures, faulted rock layers, collapsed cliffs walls and fractured flysch deposits in the shore platforms of the island. Weathering, erosion, karstification processes are evidenced within the rock layers, limestone outcrops and shore platforms of the island group of Andaman Sea.

Islands of the Bay of Bengal towards west of the main Andaman Ridge represented well developed reef terraces hill ridges, wide shore platforms and luxuriant growth of coral reefs at the shore fringes. Island interiors hills supported origin of smaller streams extended along the courses with waterfalls, springs, dry valleys, limestone basins covered by evergreen forests and littoral forests. Erosional reef terraces, limestone pavements, wider shore platforms, raised reefs, patch reefs and Ria shores are mostly fringed with the islands of oceanic environment. The wider reef flats and drowned reef platforms depicted karstic features dominated by solution channels, basins, karrens, pitted surfaces and pinnacle rocks in the Bay of Bengal fringe islands (Table 1.1).

1.4 Limestone Geomorphology in the Tropical Climate

The environmental makeup of Andaman Islands provides a significant field of weathering intensities. Carbonate rocks of the study areas show least resistance to a tensile stress and shear stress if they are compared with other igneous and metamorphic rocks of tropical climate. Areas of carbonate rocks (white clay stone, chalk, marl, mudstone, calcareous sandstone, shale limestone and milevine limestone) on the vertical cliff wall exhibit the nature of weakened tensile strength by crack development. Brittle failure within the rocks usually takes place by compressive, tensile, and shear stresses along the bedding planes, cracks and fault planes of rock bodies (Table 1.2 and Fig. 1.21).

Different rock types have different types of stresses (Bland & Rolls, 1998) and may generate varieties of surface deformations induced by tectonics (Table 1.1). As the Andaman ridge and the supported islands system are located in the active tectonic boundaries, they are prone to the impact of different stresses produced by thrust belt tectonic and seismo-tectonic in the region. Such types of stresses are reflected in the rocks of shore platforms and shore cliffs in different sections of the shore lines.

The rain water dissolved materials completely in the limestones by solution process and developed the features of caverns in underground as well as on the wall rocks of shore cliff. The green plants of tropical forests in the island interiors and concentration of algae mats and sea weeds on the shore platforms are prolific producers of carbon dioxide. The trapped carbon dioxide into the soils gives rise to greater biological activity with more production of carbon dioxide in the

Table 1.2 Different rock types with stress and stresses

Sl. No.	Rock types	Compressive stress	Tensile stress	Shear stress
1	Granite	100–250	7–25	14–50
2	Dolomite	100–350	15–35	25–60
3	Basalt	100–300	10–30	20–60
4	Quartzite	150–300	10–30	20–60
5	Sandstone (Calcareous)	20–170	4–25	8–40
6	Shale (Calcareous)	5–100	2–10	3–30
7	Limestone	30–250	5–25	10–50

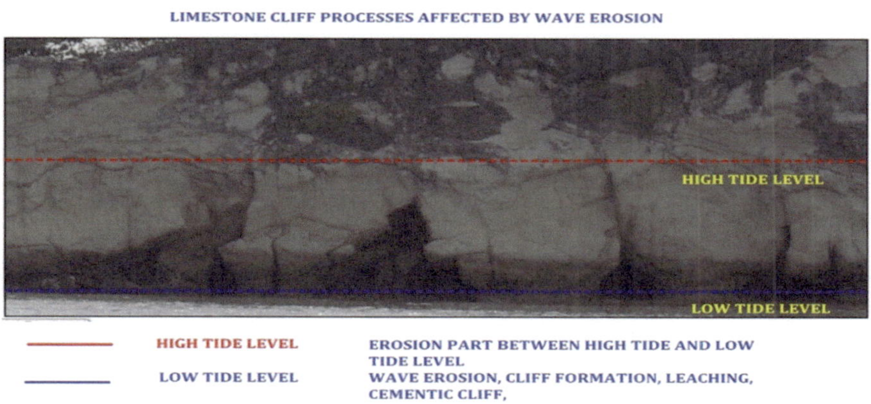

Fig. 1.21 Brittle nature of limestone cliff rocks by lateral pressure nearby Havelock Lighthouse point

region. However, the low pH (5.6) of rain water due to the presence of dissolved carbon dioxide when mixing with the limestones and limestone soils in the forested landscape that will give rise to solutional or karstic landforms.

Alternate wetting and drying also favour weathering in limestone, shale (compacted mud), marl and white clay stone as the presence of clay minerals encourage water adsorption and structural weaknesses (Fig. 1.22) of these rocks also encourage water entry. Due to their low tensile strength and inter granular pore size of rocks they are highly vulnerable to greater intensity of weathering processes in the hot and humid climate. Karstified landforms and limestone geomorphology of carbonated rocks are significant features in the island units which exhibit the uniqueness of the archipelago group in South Andaman district.

The limestones of different qualities along the rock out crop areas and shore platforms of South Andaman district revealed different types of surface deformations. The stress and strain processes induced by seismo-tectonic and thrust belt tectonics will produce different nature of surface deformations, which can provide sufficient information to identify the level of stress and strain over the exposure rocks and

Fig. 1.22 Shore platform limestones are affected by the seismo-tectonic impacts. The intense jointing, fracturing, and faulting are reflected by the stress and strain generated with active tectonic events in the shorelines of the Andaman Group of islands

morphology. The 2004 earthquake event has generated such deformation features along the shorelines of Andaman and Nicobar group of islands in the tropical oceanic coasts. The rate of weathering and erosion processes will be increased in the region after the surface deformations to modify the sediment drifting and sediment sinking processes within the boundaries of shoreline morphodynamics of the coast. Currently the morphological changes are observed after such surface deformations and by the effects of tsunamigenic hydraulic forces and normal wind waves in the islands.

1.5 Physical Settings of the Island Interior Landscapes

The islands of Little Andaman, Havelock and Neil of Archipelago formation comprise rimmed platform and ramp type platform of carbonate rocks and sediments. All three islands are fringed with coral reefs on the shelf and slope of the ramp platforms, and low tide near horizontal shore platforms. The shore platform is backed by significant cliff wall with rimmed platform facies of former sea levels. The former depositional facies of island interior platform extends over wide area of land surface in Little Andaman (Fig. 1.23).

The lithofacies, structures, deformations and sediments of the island systems indicate the role of sea level variations, tectonics and former basinal deposits in the formation of such carbonate setting. Shore platforms jointly with the coral reefs act as the carbonate machine for production and supply of organic sediments, chemogenic sediments, calcium carbonate precipitation from the sea waters and other bioclasts in the system of carbonate settings.

Topographically, the cross-sectional forms of the entire setting represent seaward fringe coral reefs, near horizontal shore platforms of intertidal areas, and island

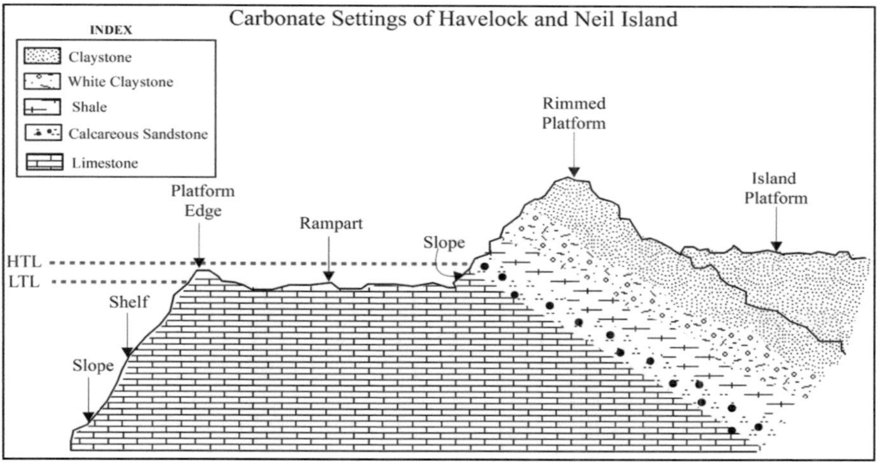

Fig. 1.23 Carbonate rocks with geomorphological features in different surface platforms in Neil Island

platform with cliffs. Sediments are derived from three different environments, such as the shallow marine environment, intertidal environment, and subaerial environment, and ultimately, they are transported and deposited into the shore platforms. Smaller tidal channels and rivers are visible in Havelock Island, but there are only two narrow runoff channels or ephemeral channels in the island system of Neil in the Andaman Sea.

However, numbers of drainage channels are extended across the island of Little Andaman in South Andaman District. The river systems fed by monsoonal rain water supply sediments from the limestone hills into the confluences over the sea board. Weathered carbonate materials as well as other siliciclastic sediments are usually transported into the modern carbonate platforms of intertidal zones from the landward side particularly during the rainy months. Another source of carbonate sediments is the fringing coral reefs of the island systems from which coralline rubbles, boulders, sands, reefal debris, residues of marine shells and other animals, algae materials, sea grasses and other bio-clasts transported into the shore platforms by waves, currents, storms and tsunamis. The shore platforms of intertidal areas accumulate sediments both from the seaward side and landward side, and sediments of such category are again reworked with actions of wave abrasion, current transport and water layer weathering. Coralline sands, gravels, dead animal bodies, debris of algal materials and siliciclastic sediments are usually transported on the HTL by wave breakers at the tidal submergence. Larger debris or materials are accumulated at the seaward margins and middle parts of the shore platform. Wave abrasion, weathering (Fig. 1.24) and erosion jointly produce granular sediments during the advancing tides onto the shore platform.

Beach materials along the HTL get cemented with supply of precipitated aragonites, calcites, calcium carbonates and dolomites with other processes of carbonate

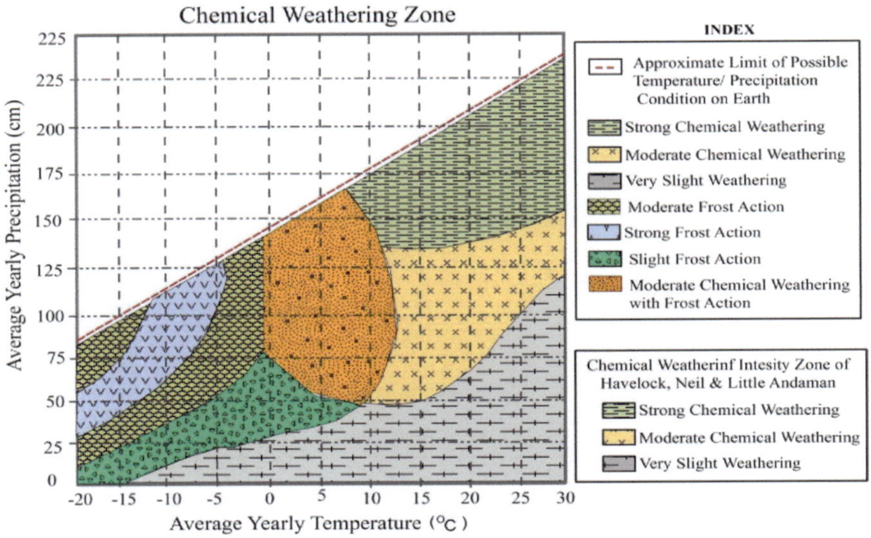

Fig. 1.24 Weathering intensity zones assessed on the temperature and rainfall data of South Andaman following Peltier's Model, 1950

degassing, and land-derived calcium bi-carbonates and fresh water seepage flow. Such type of case-hardened materials of extensive area are known as 'beach rocks' or 'Littoral concretes'. Limestone pavement surface develops on the seaward edge of shore platform with algal beds and sea grass patches by high rate of calcification process, wave beaten actions and cementation activities. Karstification processes with solution, evaporation, salt-weathering and encrustation also developed many types of karst features into the lower middle and upper parts of the modern shore platforms of the island systems (Fig. 1.25). Cliff erosion, coastal landslides, cave deposits and seepage flow waters from the fractured cliff wall supply sediments of different quality into the shore platforms.

Finally, the morphodynamics of shore platforms with interplaying of sediment supply, transport, reworking and hydrological processes by energy input, processing and energy output modify the morphology. Further, by sea level variations, sea level stabilization, shore planation and tectonics of the region, the former shore platforms have been developed into reef terraces or marine terraces along the coastal fringes of island platforms (Fig. 1.26).

The bay side beaches accumulated sediments of sand size grains with coralline debris have produced another depositional domain along the intertidal and supratidal areas by sea waves, storms waves, swell waves and tides. Cay sediments are another group of sediments which usually developed by wave deposition and wind deposition over the shore platforms and reef terraces of the island fringes. Sediments from the shelf facies are also drifted into the river mouths and back waters during the tsunami waves (2004) in the island system (Fig. 1.27).

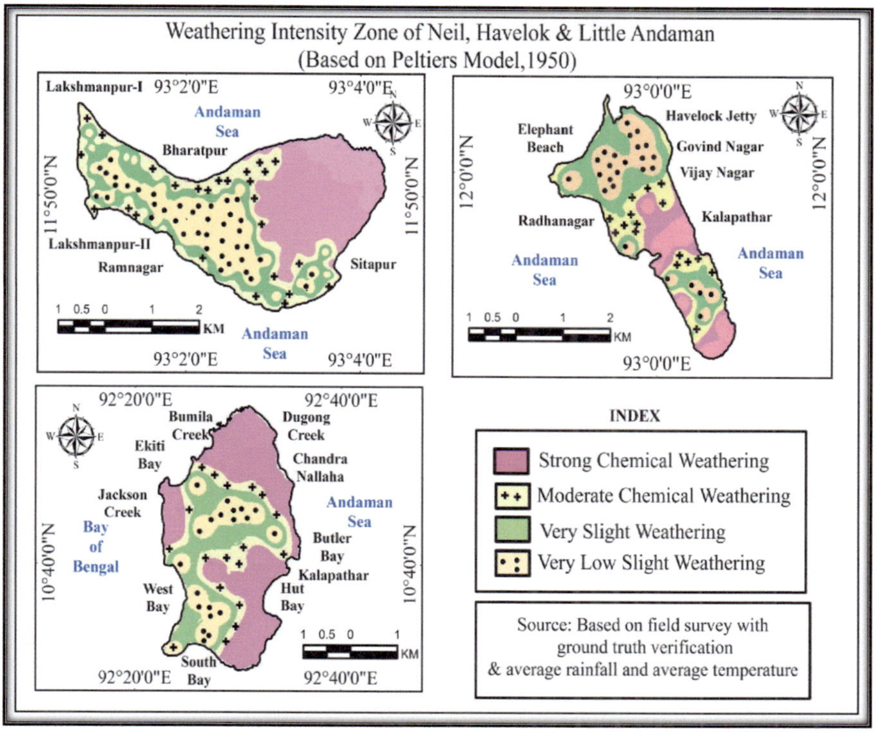

Fig. 1.25 Various rates of weathering processes in the different island units of South Andaman

Sediments sequences are exposed at the river mouths as they were affected by slow process of emergence during 2004, earthquakes in Little Andaman (Fig. 1.12). Massive and jointed limestones are found on the scarp walls of 'whisper waterfall' and also on the facies of other shore cliff of Kalapaththar in Little Andaman. Carbonate clays and finer sands are deposited in thick succession particularly around the 'Ria' shores and along the mangrove-dominated shore fringes of carbonate settings of the island towards north. Failure of the water reservoir with protective dam across the hilly streams proved the function of porous limestone at the basement of reservoir in eastern part of Little Andaman. However, the dense forests of the island protect top soils underlain by jointed limestones in the hill slopes.

Geomorphologically, the island systems of Havelock, Neil and Little Andaman are dominated by landforms of carbonate rocks and modern carbonate deposits, white clay stone, siliceous chalk and silt, calcareous mudstone, sandy limestone and carbonate limestone of Neillian formation. They are unconformably separated from the main group of Andaman Islands by Archipelago Group of Havelock Island, Neil Island and Little Andaman. Hard limestone with interbedded greenish calcareous silty mudstone and soft grey calcareous silty mudstone interbedded with

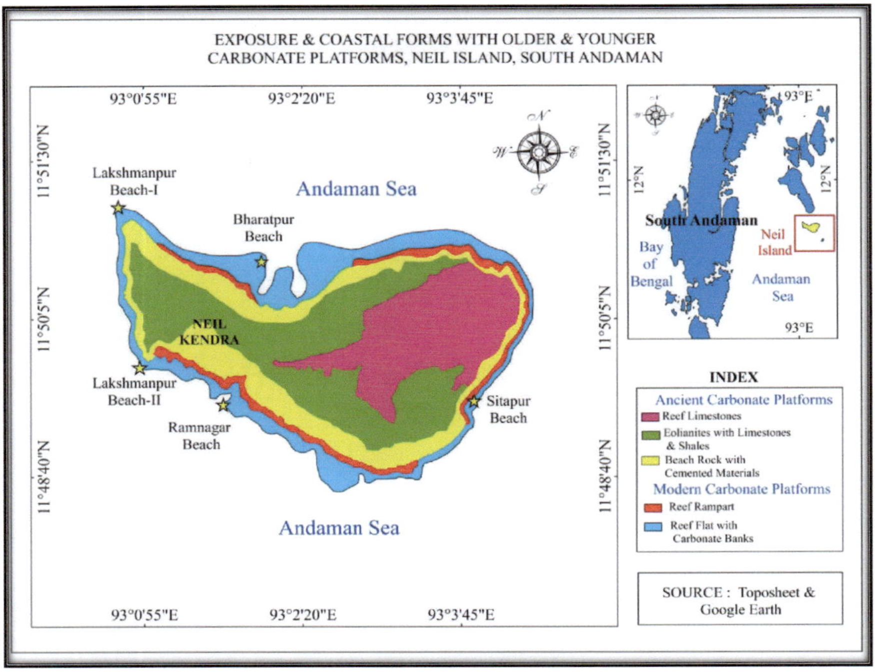

Fig. 1.26 Carbonate platforms of Neil Island

hard calcareous bands of the Havelockian stage are relatively older than the Neillian stage.

Lithologically, the cliff sections of Little Andaman contain dark grey calcareous mudstone with medium-grained calcareous sandy limestone and black colour Shelly and Coral limestone along the seashores. In many cases, the exposures of basement rocks are visible on the island fringe shore platforms and partially covered with carbonate debris, coralline sands, beach rocks and mounds of rubbles over the ramps (Fig. 1.28).

1.6 Physical Settings of the Coastal Fringe Landscapes

The island systems of the Andaman Sea and the Bay of Bengal are influenced by coastal and marine environments which can explain the genesis of the coastal features and marine features. The power of sea waves to shape the land margin is very clearly seen along the coastlines of island system in the region. Configured shorelines are modified by wave dominated erosion, transportation and deposition; changes of sea level variations over the geological time, as well as by emergence and submergence due to tectonics (Figs. 1.29, 1.30 and 1.31).

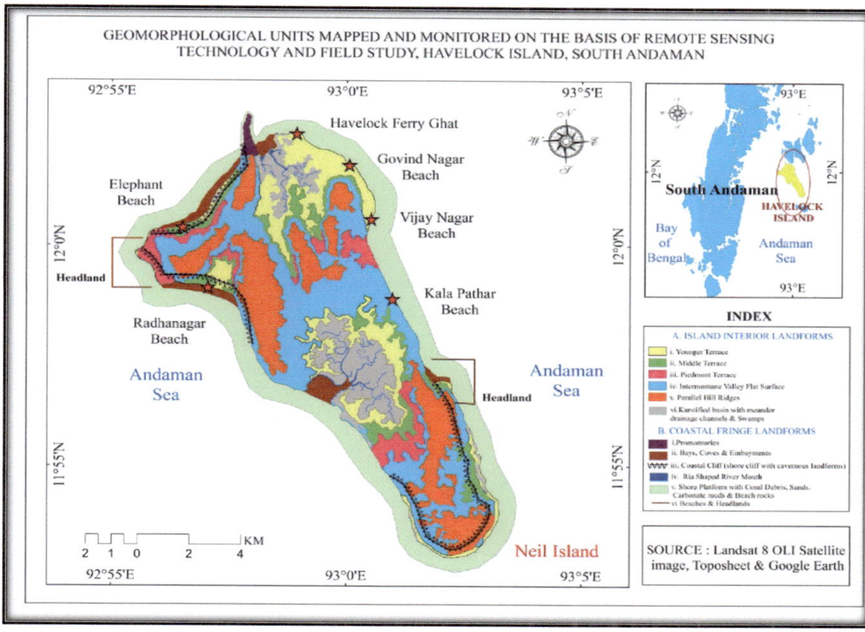

Fig. 1.27 Diversity in carbonate platforms of Havelock Island

Coastlines are re-shaped in a dynamic nature at present particularly on the windward sides of the island shores. The leeward sides on the other hand remain less active due to their sheltered locations. As the major direction of approaching waves and currents are guided by Northeast Monsoon and Southwest Monsoon drifts, the island shores are usually affected by seasonal winds and approaching waves on the both sides of the islands. Northeastern shores become active as windward side during the prevailing NE Monsoon and the opposite part of the island shores (southwest) become leeward side in the same season. Similarly, southwestern shores become active as windward side during the prevailing southwest monsoon season and the northeastern shores at the same time will act as leeward side in the region.

During the period of high intensity rainfalls, high runoff occurs along the surface of hill slopes and mostly discharges into the shore platforms and piedmont terraces. Wetlands are developed into the piedmont valleys and also into the coastal lowland basins due to the concentration of surface runoff. The recharging of ground water takes place differently on the basis of the porosity of underlying rocks and basinal topography of runoff concentration. About 156 mm rainfall took place in 12 h duration in Havelock Island and Neil Island in the month of December, 16 in 2018. Such huge rain water transported washout deposits of sands and other materials into the shore platforms, coastal basins and river valleys. It is reported that about 2 m depth of water bodies were stranded on the island surface for 3 days in Havelock and Neil Island after recharge of ground water below the sub-surface layers. Most of the fresh

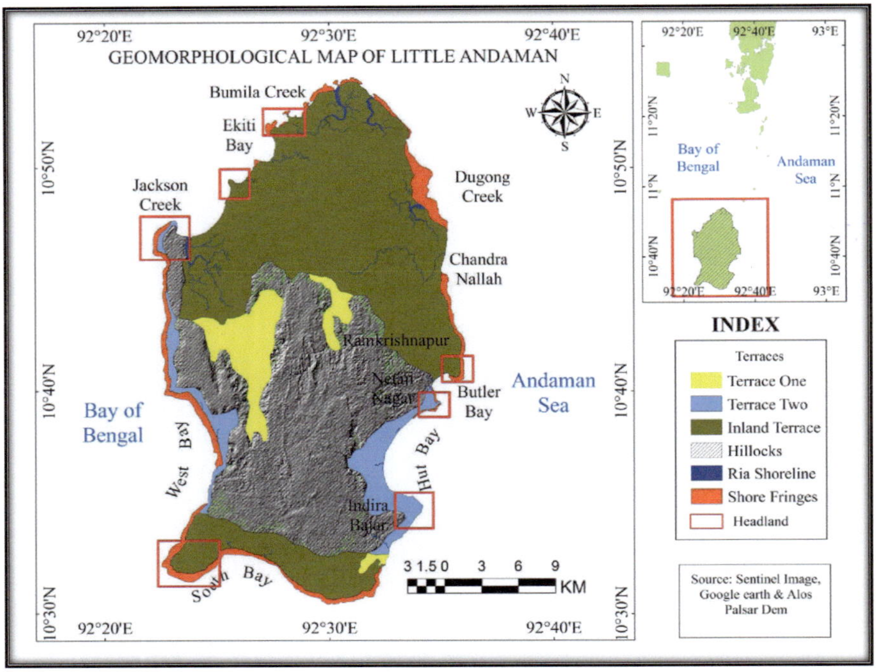

Fig. 1.28 Diversity in carbonate platforms of Little Andaman

waters usually get discharged into the sea in all the islands of Andaman group across the sloping surface in all directions. In many cases a few areas of lowland surface hold rainwaters for a few months along the coastal fringes (Figs. 1.31, 1.32 and 1.33).

Sometimes they are connected with the sea by a link channel for mixing salt waters and provide areas for salt marshes and mangrove woodlands along the coastal fringes of Havelock Island and Neil Island. However, such marshes and swamps are found into the river mouths, shore platforms and into the 'Ria' type estuaries of Little Andaman. The Dugong Creek of Little Andaman is fringed with dense and matured mangrove woodlands from inner parts to the confluence of open sea in Andaman Sea. Such 'Ria' type estuaries fringed with shallow flats provide mangrove forests to the north of Little Andaman (Fig. 1.33).

Erosion of the top soil is protected by the dense forest covers on the rolling topography and also across the hill slopes of the islands. However, the limestone-dominated rocks of jointed nature are affected by rapid weathering and chemical denudation processes, i.e. through the removal of ions in solution. The soluble calcium carbonates under the rain water and vegetated materials become effective in solution process and take part in a significant role in the chemical denudation of landscapes of such tropical climate. In many cases, the etched surfaces of jointed limestones are exposed on the seashores dominated by headlands and shore platforms of Little Andaman and

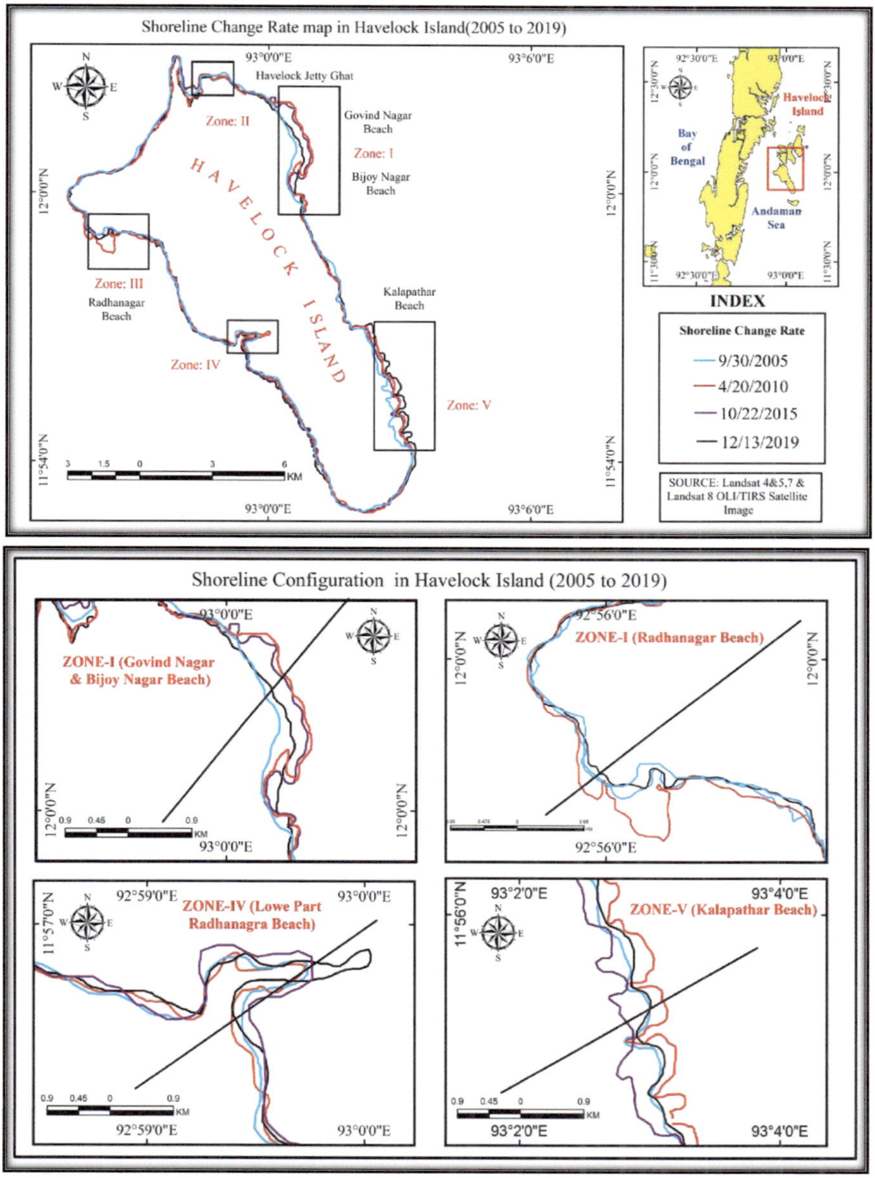

Fig. 1.29 Changing shoreline configurations in Havelock Island (2005–2019)

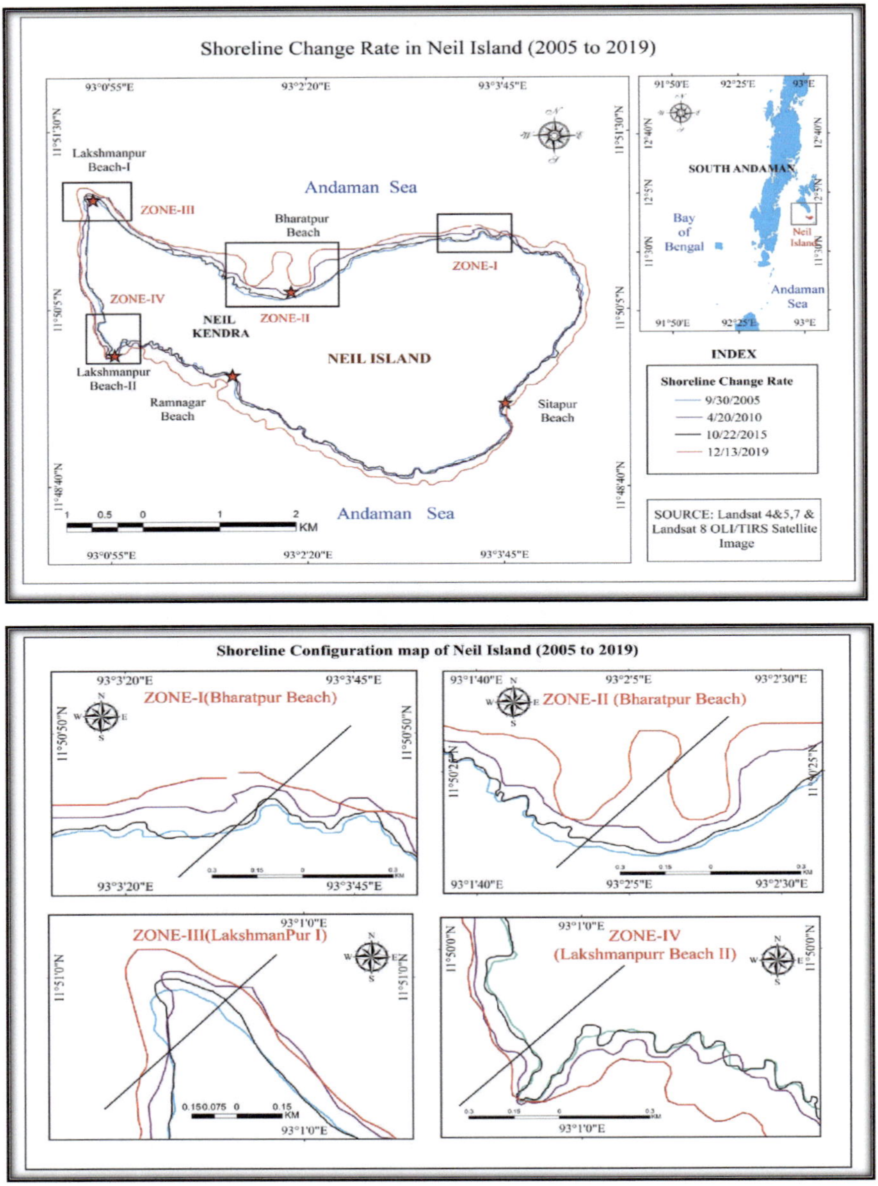

Fig. 1.30 Shoreline changes in Neil Island (2005–2019)

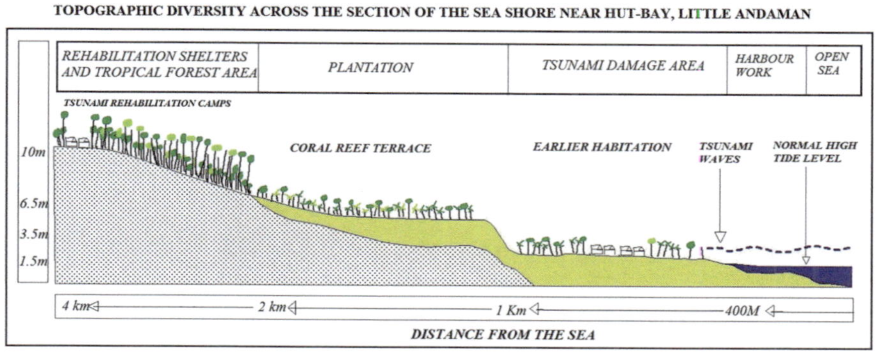

Fig. 1.31 Reef terraces and shifted rehabilitation centres on the higher grounds in the sea face of Hut Bay, Little Andaman

Havelock Island. The marine environments of these oceanic islands played a dominant role in shore planation process by swell waves, seasonal wind waves, storm waves, tsunami waves under ephemeral rise of sea waters and by sea level variations through the Pleistocene-Holocene climate changes. Open marine environment therefore provides condition for natural wave attacks from all sides over the island shorelines across the ocean surface based on the approaching winds.

The shoreline configurations are modified over the time scales particularly due to the local rock structures and lithology's, openness of the shorelines to approaching wind directions, concentrated wave energies rate of chemical denudations, and dynamic etchplanations. Coastal habitats like fringing coral reefs, algal reefs, vermetid shell ridges, sea grass patches and mangroves along the shore platforms and sub-tidal banks however, minimized the rates of erosion and sediment transportation under wave energies during normal conditions.

Fringing coral reefs, shore platforms and shore cliffs of Andaman Islands (Fig. 1.34) are partially damaged due to the occurrences of 2004 earthquakes (26th December) and tsunami generated wave attacks (Figs. 1.33 and 1.34). Raised coral reefs are recorded on the shore fringes of the Bay of Bengal along North and Middle Andaman Islands (i.e. Interview Island, Landfall Island, etc.) and submerged coral reefs are found on the shore fringes of South Andaman group of islands after the action of emergent and submergent processes respectively during the events of seismo-tectonic activities in 2004. Coral reefs of Havelock Island and Neil Island were fractured and subsided during the event of earthquake (9.1 Richter scale). In few places of Harmindar Bay, Hut Bay, and Buttler Bay of Little Andaman, the shore platforms experienced upliftment. The oceans are also heating up at faster rate in previous years than ever before and causing extreme weather events like cyclones, high-intensity rainfalls and marine heat waves in the region. The coral bleaching and stronger cyclones resulted from such effects of warming seas at present.

Tsunami waves of 2004 drifted coral rubbles, coralline boulders and 'cay sediments' on the shore platforms. Marine sands, shingles and debris of coral fragments

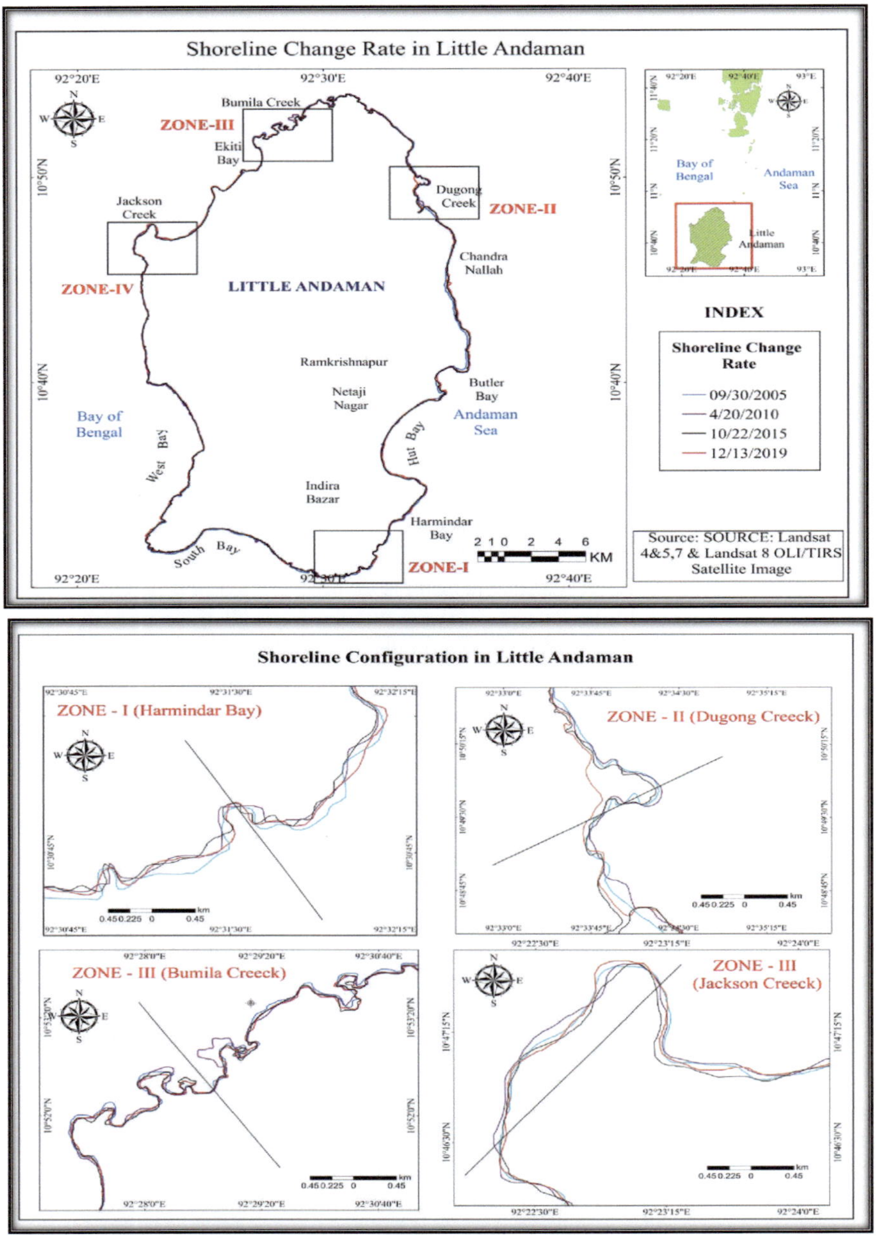

Fig. 1.32 Shoreline configuration dynamics in Little Andaman (2005–2019)

Fig. 1.33 Shore platform morphologic features (**a–d**) evolved under the marine abrasive forces by the sea waves

Fig. 1.34 Coastal cliffs developed on the up-thrust plane of the rock bodies (Havelock Island)

and oyster shells are deposited into the river mouths or into the submerged estuaries of the region. The sands and debris of graveliferous corals are encroached into the reef terraces by wash over fan deposits in the stronger cyclones at and around the island fringes of Little Andaman, Havelock and Neil Island.

1.7 Neo-tectonics and Seismo-Tectonic

Forests were affected by upliftment and subsidence of the coastal fringes due to seismo-tectonic activities in 2004 in the islands. Such vertical movement of the land surface provided surface instability along the crestal part of the shore cliffs from which large trees uprooted and collapsed on the sea face. On the western shores of Havelock, Neil and Little Andaman Islands, many trees are horizontally lying across the shore platforms and uprooted from the top soils due to the impact of seismo-tectonics in the region (2004).

The seismic waves generated by seismic event of the previous times have produced several morphologic signatures of upliftment and subsidence phenomena in the shore fringe areas. Researchers have estimated the rate of upliftment compared with historical seismic events and their date of occurrences following the reef terraces along the shore fringe areas of Andaman group of Islands (Paul et al., 2012; Rajendran et al., 2008; Rajendran & Rajendran, 2022). By such estimation it is observed that the upliftment rate of reef terraces at every seismic event were ranging from 0.93 to 1.66 mm/yr in the Andaman Group of Islands. Thus, counting the location, altitude and distance of the reef terraces from the high tide shoreline, the events of seismo-tectonics may be estimated over the island landscapes. There are four to seven (4–7 nos) numbers of reef terraces and marine terraces in the oceanic coasts which are visible along the Bay of Bengal fringe shorelines of main Andaman ridges, and indicate the occurrences of historical mega seismic-events (Fig. 1.35). Usually, the seismo-tectonic impacts also generated energy levels after the earthquake events to produce ruptures, fissures, jointing, tilting within the rocks and reactivation of faulting in the archipelago in the geological past during Holocene-Pleistocene epochs.

1.8 Marine Processes and Their Impacts

1.8.1 Tsunami (2004)

After vertical movement of the land surface along the sea face, tsunami waves pushed inlands with salt waters, marine sediments over the subsidence prone forested landscapes to produce submerged forests in the islands. Large tracts of forests were

Fig. 1.35 Location of mangrove swampy forest over the raised reef terraces of limestone pavements on the bank margins of Dugong Creek in Little Andaman

submerged by the blanket of sediments after the event of subsidence into the embayments and estuaries in 2004. Mangrove forests in Little Andaman and Havelock Island shifted into supratidal surface due to upliftment by event of 2004 earthquake.

1.8.2 Cyclones

A few cyclonic storms (2012, 2013, 2015, 2016, 2018 & 2020) with surging waves and torrential rains inundated coastal fringes in South Andaman Islands. The uprooted trees were drifted along the shores by strong longshore currents generated in the events of marine inundations (Table 1.3, Figs. 1.36 and 1.37).

Field studies indicated the presence of drifted wood logs over the shore platforms in all three islands. Log drifts are major phenomenon in the carbonated islands due to the occurrences of large-scale marine inundations. Such events also supplied external sediments of inorganic characters into the shore platforms in which mangroves accumulated blanket of sand size sediments. The pneumatophores of mangroves were trapped by such inorganic sediments and for which the standing mangroves were damaged largely on the shore platforms.

1.8.3 Waves

The reef flats, reef banks and shore platforms are directly exposed to the wave attracts in the oceanic islands. Sea waves are generated by wind speeds, wind directions and fetch distance in the oceanic environment and when they will be propagated over a long distance across the sea surface they are known as swell waves. Strong wind waves generate at the time of atmospheric low pressures and cyclone land falls at the coast. Open marine environment however, generate also significant waves at the windward side of the island shores to produce impacts on the shore platforms. High

Table 1.3 Frequency distribution of tropical cyclones in the Bay of Bengal between 2007 and 2020

Sl. No.	Cyclone	Landfall	Wind speed (km/hr)	Pressure (mba)
1	Sidr	16-11-2007	225	944
2	Nargis	03-05-2008	185	960
3	Aila	25-05-2009	180	968
4	Phyan	11-11-2009	90	988
5	Nilam	01-11-2012	100	990
6	Phailin	13-10-2013	215	936
7	Halen	19-11-2013	130	990
8	Lehar	28-11-2013	140	980
9	Hudhud	14-10-2014	215	950
10	Vardah	13-12-2016	165	962
11	Maarutha	17-04-2017	80	996
12	Mora	31-05-2017	110	978
13	Fani	05-05-2019	215	932
14	Bulbul	12-11-2019	190	976
15	Amphan	21-05-2020	240	928
16	Yaas	26-05-2021	165	972

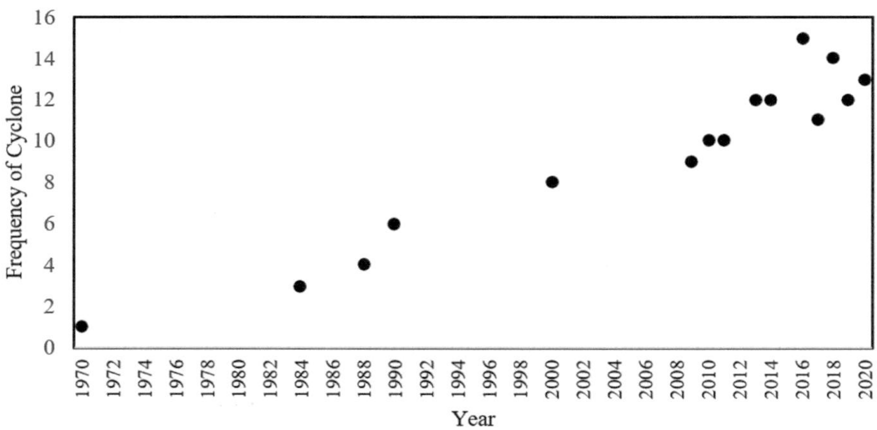

Fig. 1.36 Showing the concentration of high magnitude cyclone frequency in the previous decades (1970–2020)

tides on the other hand pushed the wave breakers into the foot of the cliff walls to cut the basement by wave energy along the high tide shorelines (Fig. 1.38). Cliffs are notched in island fringes which represented the level of wave movement at the surging tides (Fig. 1.39).

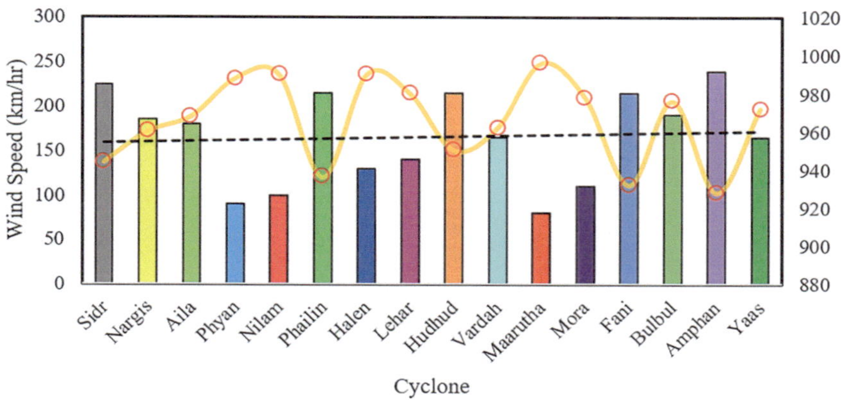

Fig. 1.37 Bar graphs representing the wind speed and atmospheric pressure of the previous cyclones

Fig. 1.38 Hore platforms of Havelock Island and Ross Island modified by waves and tides

Fig. 1.39 Shell limestone platform of Kalapathar (Havelock Island) is affected by notching and caving by the wave cut process at the high tides

The recession of reef banks by waver breakers have created buttressed banks with reflective wave energy at the low tide shore lines of the wind ward reef front position. Features like glow holes, undercutting tunnels and caving developed on the underwater reef bank walls by wave breakers along the low tide shorelines of coral fringed coasts. Wave pounding effects at the rising tides have produced cemented carbonate crust on the lithothamnian flats where the brown algae mats and green algae mats covered the shore platform edges in South Andaman Islands. Wave abrasion, water level weathering and salt weathering processes produced surface etching of the shore platform and transported sediments into the morphodynamics systems and their subsystems of the shore platforms at the mid tides and high tides along the shore fringe environment (Fig. 1.39).

The interpolated wave data at the regional scale originally generated by INCOIS wave Buoy station data of a few places of the Bay of Bengal shores and Andaman Sea shores indicated the spatiotemporal variation of the waves' heights and approaching angles. Height of the sea waves during the south west monsoon ranges between 1.5 and 2.5 m on the Bay of Bengal fringed coasts, whereas it ranges from 1.5 to 2.0 m at the northeast monsoon. During atmosphere depression and cyclonic storms, the wave height increased up to 4.0–6.0 m at the windward site of the oceanic island. Such great waves contributed major roles in sediments supplies and sediment transport across the shore and along the shore of the coral fringed coasts (Fig. 1.39).

1.8.4 Tides

Tide waters submerged the fringing shore platforms around the islands twice daily and pushed the wave breakers into the high tide lines to run the cross-shore process in the oceanic islands. Tidal submergence usually adds surface and subsurface moisture into the shore fringe gently sloping platforms, as well as restore the tidal life of rocky shores and pools of salt water bodies in the shore fringe environment. Various types of algal growth, patch reefs, sea weed and sea grasses extended their colony into the shore platforms which are usually fed by temporal submergence of tide waters. During submergence, sediment transportation, surface etching and wave abrasion become active in the tidal zones, and tide water deposits carbonate, aragonites and other dissolved compounds, ions and nutrients that organisms can use for energy. The calcification ion process also run by the temporal submergence of tides in the shore fringe platforms of the islands in the high tides.

Similarly, the subaerial exposures during low tides provide condition form evaporation and karstification process in the shore platforms by active chemical and bio-chemical weathering mechanisms in the tropical coast with addition of rainwaters and mixing of dissolved carbonic acid breached by seepage waters at the shore platforms from the island fringe environment. The tidal elevation ranges from 0.90 to 1.00 m but occupies a large area of the gently sloping shore platforms during high tides. The tidal height also ranges significantly in the embayed shores of the islands during summer months. Ria shores dominated by estuaries also behave like arms of the seas during the rising tides in the tropical oceanic islands. Mangroves of the shore platforms and back shores also receive the supply of carbonate sediments and nutrients in the high tides. The seed dispersal and regeneration of mangroves are also guided by high tide in the open marine environment.

1.8.5 Winds

Winds play a very constructive role in the windward sides of the reef building islands by generating the waves heights and approaching directions. The direction and velocity of the winds vary in the seasons during the atmospheric depression. The Andaman ridge with archipelago is extended from north to south up to 1200 km long barrier which separated by the Andaman Sea from the Bay of Bengal. However, the northeast monsoon winds prevail in the winter months and southwest monsoon winds occur in the summer months in the region across the long fetch distance over the seas. Thus, a single island of the archipelago gets wind velocities and directions and feel the effects of windward sides and leeward sides in two different seasons. The southwestern shores receive high wind velocities in the southwest monsoon but the northeastern shores receive high wind velocities during the winter months of northeast monsoon winds. Thus, the luxuriant coral reefs are found in the windward sides of the islands. During cyclonic weather, the wind velocity oscillates from 65 and 120 km/hr in the oceanic islands. Normal winds vary from 30 to 50 km/hr in the dry season but able to produce strong wind waves on the windward sides of the islands.

1.9 Physical Responses of the Coral Fringe Karstified Islands to the Indian Ocean Tsunami (2004)

The great earthquake of Andaman (2004) produced a large-scale impact with tsunami genic forces over the shore fringe areas of the Andaman and Nicobar group of islands (Paul, 2005; Acharyya & Dasgupta, 2006). Both the earthquake and tsunami effects are reflected on the rock outcrops, shore platforms, shore cliffs, shoreline morphology, reef flats, reef banks, island platforms, estuaries, promontories, reef terraces and intertidal flats of the island systems from north Andaman to southernmost tip of Great Nicobar Island.

Upliftments, subsidence, ruptures, fractures, fissures, jointing, tilting, faulting, collapsing, with generated stresses and strains produced several deformations over the surface and rock out crops due to the sudden effects of such seismo-tectonic event in the Island system. Later, the tsunami-induced hydraulic forces of the sea water have driven the energy to drift the sediment materials from the sea bed, to erode the sediments from the shorelines, and to deposit the sediment into the basins of the archipelago (Figs. 1.40, 1.41 and 1.42).

The coral reefs of North Andaman, Middle Andaman and South Andaman on the Bay of Bengal fringe shores revealed upliftment which are well reflected in Interview

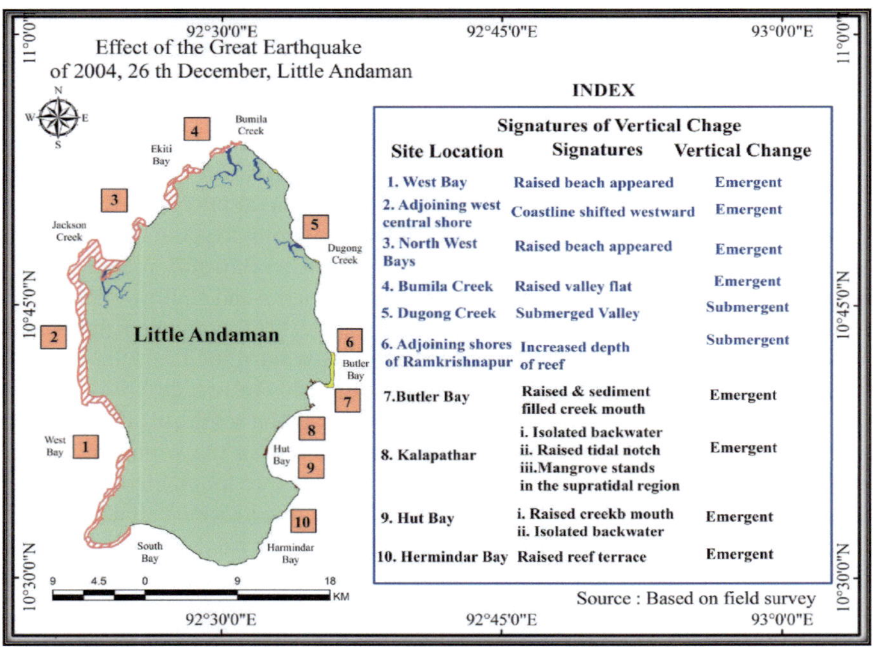

Fig. 1.40 Vertical changes in shoreline features after the occurrence of the great Earthquake (2004) in Little Andaman

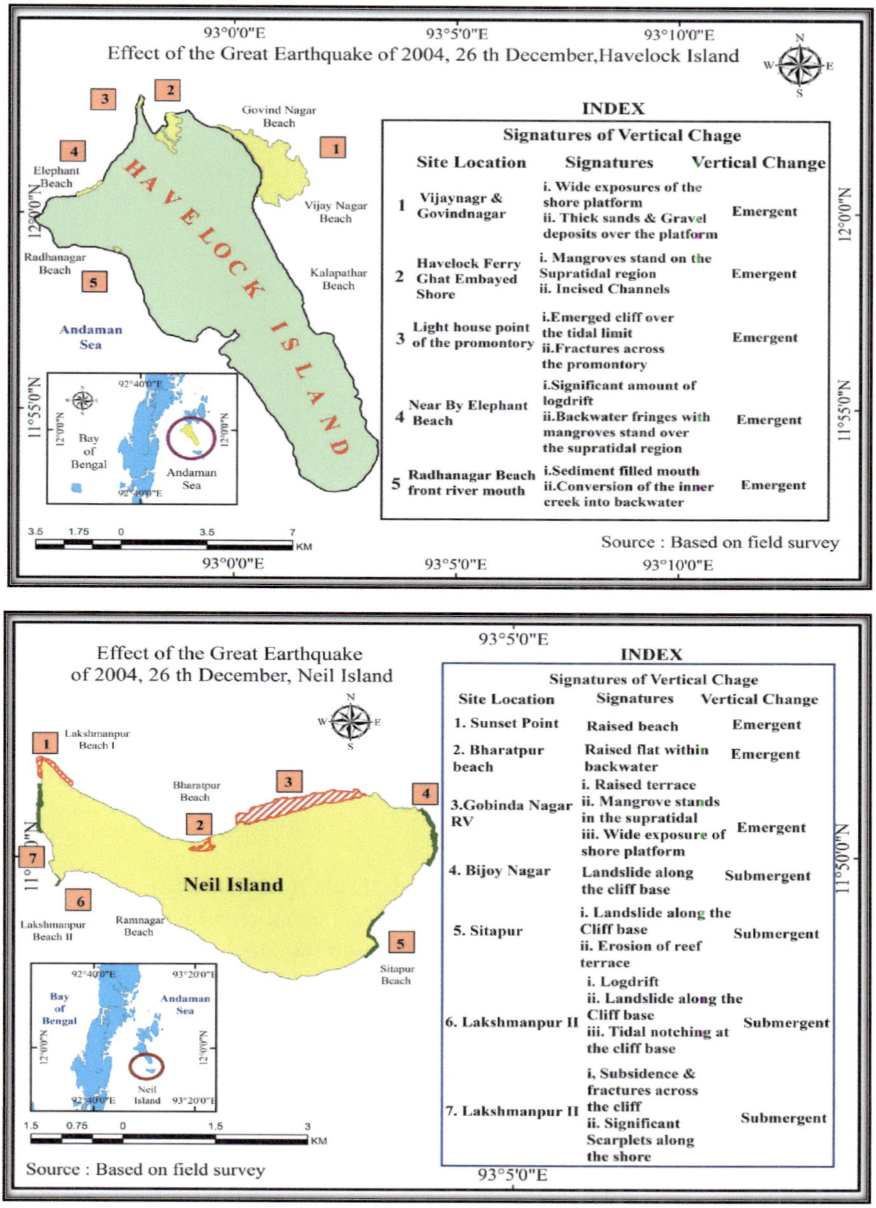

Fig. 1.41 Diversity of vertical changes in the shorelines of Havelock and Neil Island after the 2004 earthquake

Fig. 1.42 Vertical changes in the shoreline and river mouths in Little Andaman and Havelock Island due to the impact of seismo-tectonic upliftment

Island, North reef Island, North Sentinel Islands, Little Andaman and red skin Island. As the reef flats are uplifted and exposed to the subaerial environments, the active weathering processes with erosion and lithification modified the morphology and converted them into reef terraces for addition of coastal labs. In future there will be the formation of eroded reef terraces by oscillation of the sea level with advancing wave breakers into the shores. If the sub-tidal zones become intertidal after the event of upliftment, the previous intertidal areas also will be converted into supratidal zones by the similar effects (Figs. 1.42 and 1.43). Thus, the intertidal mangrove habitats of the shore fringe areas and island interior linear valley flats are lying over the supratidal fringe keeping isolated themselves from the active shorelines at present. A few mangroves can survive in the supratidal zones to avoid the physiological stress by extending their root into the shores but other sensitive tidal wood lands could not exist in the uplifted zones of tidal flats of the entire island system.

On the other hand, the subsidence-induced reef flats, estuaries and patch reef of the shore platforms are alarmingly affected by the sedimentation, boulder drifts and energy tresses in the island shores. Zooxanthalese and coral were declined in

Fig. 1.43 Images showing the formation of 'Ria' shores due to land subsidence at Katchal Island (Nicobar Group) after the event of the Great Andaman Earthquake, and consequently, the drifted logs and sediment filling processes were the most important activities during the tsunami event in the Indian Ocean region

Havelock Island, Neil Island, Jolly Buoy Island and other islands of South Andaman due to such events of earthquake and tsunami forces. The reef flats were fractured by seismo-tectonic forces in parts of Havelock Island and Ross Islands. By give up and catch-up processes the corals can adjust with time to maintain their growths but rates of growths may not adjust with the oscillations of sea level in the marine environment (Table 1.4, Fig. 1.43).

The limestone pavements with karstic features were affected by cover deposits of sediments after land subsidence in places whereas the cover deposits have removed from the limestone pavements after slight upliftment in other places of the island shores of little Andaman and Neil Island. Reef terraces of Wandoor, Chidiya Tapu, Khurmadera, Harmindar Bay, Sitapur, Ramnagar sections are rapidly greeting eroded after the slight rate of subsidence and wave abrasion forces. Many fissure caves were expanded in the promontories of the embayed shores by the impacts of seismo-tectonics and rapid weathering processes through solution breaches of rainwaters in the archipelago group of geological formations.

Significant output of boulder drifts and log drifts have occupied the shore fringe surface depressions in the estuaries, embayments and shore platforms of the Nicobar group of island and South Andaman Islands after the event of tsunami (200). The shore cliffs were collapsed and fragmented, but rock slides and debris fall occurred from the higher crest of the cliffs into the foot slopes of the vertical walls after the seismic events in the regional tectonic settings. Sitapur cliffs, Lakshmanpur cliffs, Nilian west cliffs, Radhanagar cliffs, and Kalapathar cliffs of the South Andaman Islands represented mass wasting features after the event of great earthquake. However, the boulders and debris surface have been modified by tafoniferous activities with wave splashings in the high tide's levels and also by rainwater solution process activated in the monsoon months in the region.

Table 1.4 Events of Andaman earthquake and Indian Ocean tsunami (2004) produced remarkable effects along the shore fringe environment

<table>
<tr><th>Events of seismo-tectonic activities</th><th>Effects of earthquake</th><th>Effects of tsunami</th><th>Remarks</th></tr>
<tr><td>The Great Andaman Earthquake,2004</td><td rowspan="5">(1) Vertical displacement of the shorelines (2) Ruptures, fissures, fractures, jointing within the rocks (3) Faulting in the shore platforms and stratigraphy (4) Collapsing the shore cliffs (5) Mass wasting in the shore cliffs (6) Reef terraces by vertical upliftment along the coral fringe shores (7) Ria shores by vertical land subsidence in the estuaries (8) Surface deformations across the roads (9) River mouth raised</td><td rowspan="5">(1) Sediment drifts (2) Boulder drifts (3) Submerged corals (4) Log drifts and sediment filling into the basin (5) New areas of Inundations (6) Forest damages in the shore platforms (7) Coral damages in the shore fringe areas of islands (8) Ports and Harbour affected by waves (9) Settlements displaced from the low lying shores</td><td>Erosion of reef terraces increased rapidly in South Andaman Islands</td></tr>
<tr><td>The tsunami event of Indian Ocean, 2004, 26th December</td><td>River systems affected significantly</td></tr>
<tr><td>Focus of the earthquake was at the seabed of N.W. Sumatra Island in Indian Ocean (32 km depth)</td><td>Raised platforms</td></tr>
<tr><td>Magnitude of the earthquake was 9.2 RH</td><td>Islands fractured</td></tr>
<tr><td>Maximum tsunami wave height was > 10 m above the surface</td><td>Supply of carbonate sediments into the shorelines Lighthouses separated</td></tr>
</table>

Chapter 2
Coastal Geomorphology and Habitats

2.1 Coastal Features with Regional Landscape Diversity

Regionally, the island landscapes are diverse in physiography, terrain units, composition of the carbonate platforms, cliff morphology, stratigraphic exposures and age-old reef terraces. In the next section of the thesis the diversity is reflected in the following ways.

2.1.1 Physiographic Diversity of the Islands

Physiographically, the islands are classified into various types of landform assemblages on the basis of SAR data, ALOS PALSAR DEM (ASF) and extensive ground truth verification on temporal scale and spatial scale. Considering the contour patterns prepared from the ALOS DEM for above-mentioned islands and with repeated field visit for monitoring geomorphology in the islands, several physiographic units are classified to discuss the geomorphological characters of landform assemblages (Fig. 2.1).

The island surface hills and valley flats with siliceous chalks occupy high relief (30–10 mts = 20 mts) in the entire island. Dissected hills are separated by narrow valley flats and they are drained by ephemeral drainage channels. The hills with siliceous chalks are joined, porous and fractured but covered by dense forests of evergreen trees. Denuded hills with surface runoff and chemical weathering intensity during monsoon month's supply sediments into the valley flats. Thus, gradual valley fill process provided higher valley flats between the hill ridges in the region. Highland fringes are crenulated and marked by significant scarps, piedmont slopes and piedmont terraces.

Inland fringed reef terraces with calcareous sandy limestones and coralline limestones have descended towards the shore cliffs in stepped surfaces. Raised reef

A. Paul et al., *Karstified Coasts of the South Andaman Islands*,
https://doi.org/10.1007/978-3-032-10299-7_2

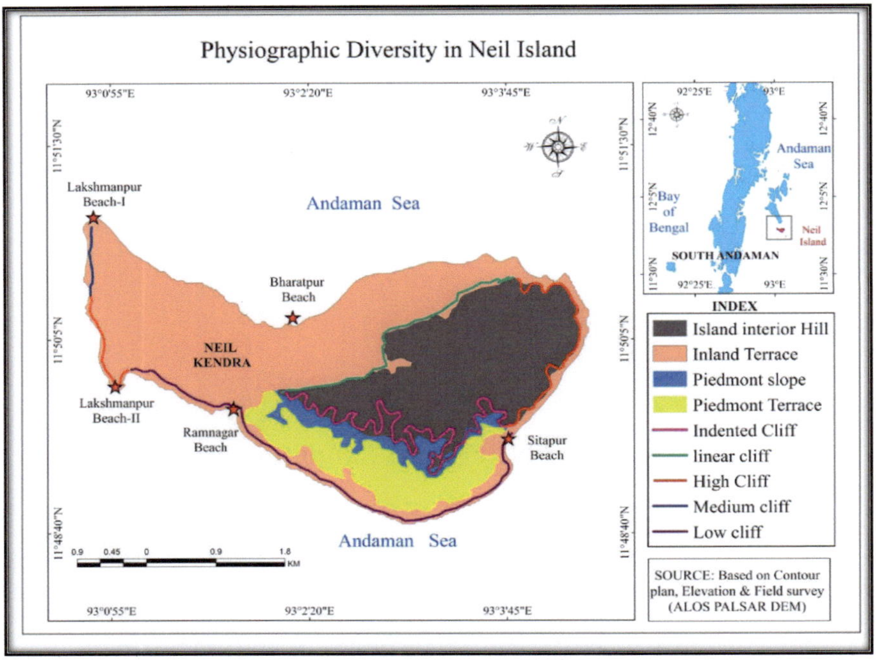

Fig. 2.1 Physiographic units of Neil Island

terraces along the shore fringes of northern parts of the island are sparsely vegetated and relatively lower parts are occupied by mangrove wetlands. Limestone soils of the surface are cultivated and occupied by human settlements. The porous limestones are recharged with monsoon waters in the rainy months. The soils are sandy and graveliferous with coralline debris.

The shore cliff sections with bio-calcarenites, white clay stones, muddy limestones and silt stones provide ideal sedimentary facies and stratigraphic layers. They are unconformably lying over the calcareous sandstones and massive limestones. There are five cliff sections with fossiliferous limestones of different categories, i.e. high cliffs, low cliffs, indented cliffs, linear cliffs and medium cliffs along the island fringes. Significant cliffs are found on towards south, west, southeast and east of the island rims with 23–12 mts elevations, and fringed with near horizontal shore platforms.

Exposures of shore platforms on the sea-face with apron of coralline sands, coralline rubbles, beach rocks and algal mats that occupy the most dynamics surface morphometry under tidal, sub-tidal and marine environments. Carbonate clays, mangrove swamp clays, boulders and other granular materials derived with mass wasting process at the cliff base and soluble limestones washed into the shore fringes areas are included as recent origin. Coral rags on the shore platforms are cemented and covered by thick algal mats. Karstified features (caves, taffoni, pinnacles, lappies, etc.) are found along the transitional zone between the vertical cliff walls and shore platforms. Weathering, wave abrasion and tidal current scours with water layer weathering modify the shore platforms surface morphometry on the windward sides. 'Sediment cays' or 'Sand cays' are located on the leeward sides of concave banks of the island, and they are generally transported by storm surges, swell waves and tsunami waves.

Neil Island is characterized by the location of an elevated island surface hilly terrain with valley flats particularly composed of siliceous chalk on the eastern margin. The hill ridges of 30, 25 and 20 m heights are extended into stepped surfaces and dissected by encroached valley flats towards the southeast direction. The lower stepped surface of 20 m elevation towards south and southeast direction is truncated by narrow valleys with ephemeral channels. Significant scarp walls exist on the valley margins of encroached flats. There are some isolated uplands of 10–15 m elevations on the western fringe of Neil Island. Island fringe reef terraces (3.00, 5.00 and 10.00 mts in heights) with calcareous sandy limestones and coralline limestones occupy not only the coastal margins, but also extended over the interior parts. This part of island surface gets inundated by high-intensity rainfall (167 mm on 7th December, 2016) and also during the ephemeral rise of sea water in the occurrences of stronger cyclones or storms (Vardah-13th December, 2016; Lehar-28th November, 2013; and Pabuk-7th January, 2019) over Andaman Sea. There are two surface depressions nearby the sites of Bharatpur and Neil Kendra in which the mangrove wetlands exist with brackish water environment. Two other sites of Sitapur Beach and Ramnagar Beach are drained by two smaller channels of ephemeral character (Fig. 2.1).

The shore cliff sections with bio-calcarenite, white clay stones, muddy limestones and siltstones are significant geomorphic unit along the shore fringes of Neil Island. On the sea face of Sitapur beach such cliff section is elevated up to 20 m with exposures of white clay stones and chalks. Sub-vertical to near vertical joints and angular fractures have shattered the rocks of cliff sections, and for which rock falls and pseudo-karstic caves with seepage waters across the walls in the rainy months modify the basal slope of the region. Similar cliff sections are found in the shoreline of Lakshmanpur-II in which the layers of graveliferous limestones, sheety limestones, massive limestones and calcareous sandy limestones are exposed on the vertical walls. Wave erosion, wave splashing, weathering, seepage pits and jointing have played significant role in modification of cliff morphology with resulting features of notching, caving, arching and phytokarstic group (Fig. 2.1).

Bio-calcarenites or the fossiliferous materials with calcareous sandstones, limestones, laminated shales and shelly limestones are exposed on the shore cliff sections of western margin of the island. This cliff section is affected by fractures, jointing, subsidence and slumping after the 2004 earthquakes. Number of marine animal fossils, coral fossils and algal mats of decomposed particles are visible on the debris of cliff rocks and also on the layered structure of cliff walls.

The island adjacent of Richie's Archipelago is composed of round chalk with white clay stone, Melville limestone, Muralat Jig chalk and white clay stone and they are unconformably blanketed with raised beach materials, mangrove swamp clay deposits, alluvial and colluvium materials. Shell limestone, sandstone, clay stone of

Nicobar Group are separated by Archipelago group and Quaternary Holocene group of formation in the island. Most of the hills of Havelock Island are formed with chalk and white clay stone layers and they are affected by jointing, thrusting, tilting and faulting at the sea face. Exposures of fault lines across the shore platforms and thrusted layers on the vertical shore cliffs represent the presence of stress field in the island (Fig. 2.2).

There are two groups of drainage channels within the island landscape. Channels of the intermontane valley flats in parallel patterns aligned with the NW to SE direction, and the torrents or swift flowing streams across the coastal hills on the seaward slopes with presence of knick points and waterfalls belong to such groups. The parallel hill ridges are elevated up to 30 mts and separated by intervened depressions (<5.0 mts). Centripetal drainage pattern is developed into the depression of Havelock Island across 11° 55′ N latitude. Significant scarps are found on the hill slopes toward western shores and also at the northeast facing parallel ridges in the island (Fig. 2.3).

Topographically, the landform units of Havelock Island are categorized into: hill ridges with sandy limestone and white clay stone; pediment slopes with hill front deposits; intermontane valley flats and wetlands with wash deposits; shore cliff sections with clay stones, mudstones, sandstones and siltstones; reef terraces with apron of sands and coralline limestones; exposures of shore platforms with limestones and phytokarstic features; bays, caves, promontories, beaches with thick sands and coralline debris and beach rocks. Such diversity of landforms has resulted from chemical weathering, denudation process, water layer weathering, wave abrasion, surface runoff concentration, drainage activities and karstification processes. Both the low-level terrace and high-level terrace with sands, coral debris and other alluviums indicate the effects of neotectonics on the island platforms. They are also named as the raised beach features and reef terraces in the islands of archipelago group. Denudation and pedimentation processes have modified the original hill ridges

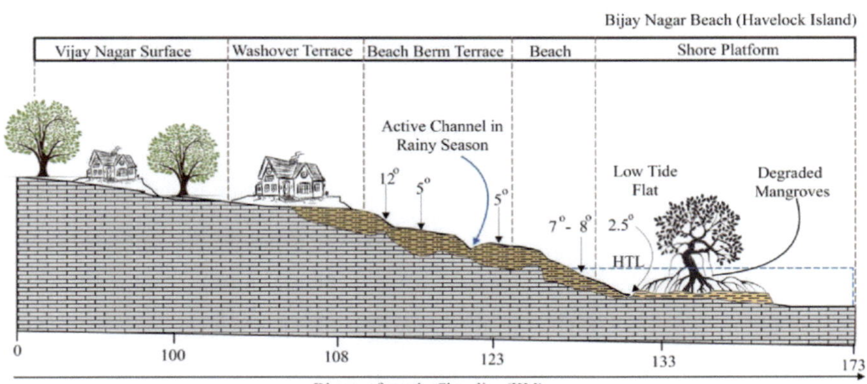

Fig. 2.2 Shore face terrain units at Bijay Nagar section of Havelock Island

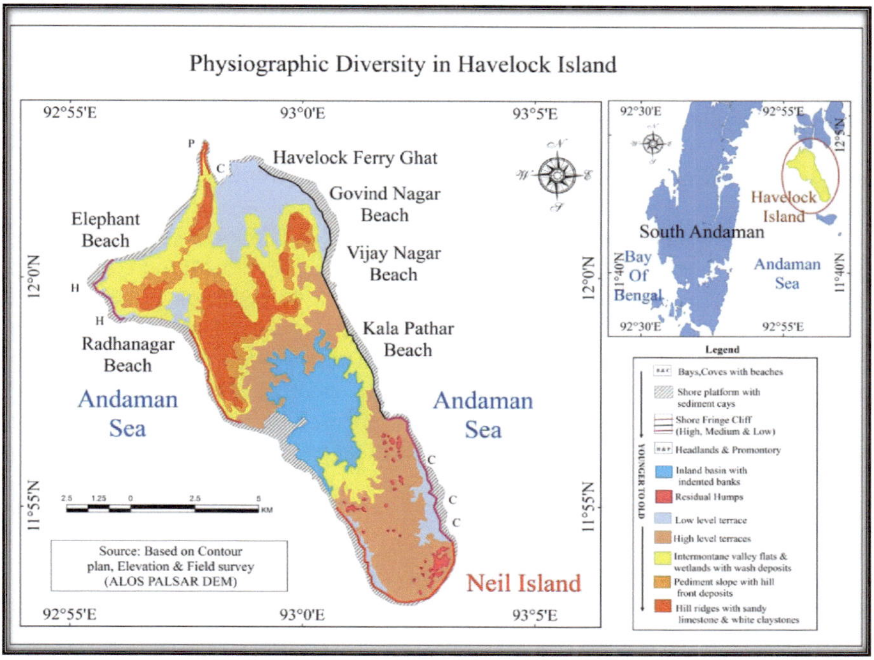

Fig. 2.3 Diversity in physiographic units, Havelock Island

into denuded hills and piedmont surfaces. Cavernous sediments, mangrove clays and other sediments are deposited in the shore fringe tracts of the island particularly along the bays, coves and depressions (Fig. 2.4).

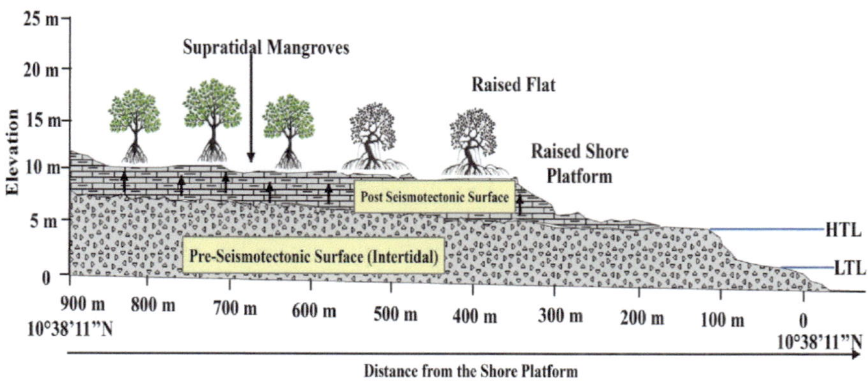

Fig. 2.4 Pre-earthquake intertidal flat uplifted into supratidal flat due to seismo-tectonic activities during 2004 on the eastern shores near "Kalapaththar" (Little Andaman)

About 86% areas of the island are densely covered by tropical evergreen forests, monsoon forests and mangrove forests. Calcium bicarbonates with soluble limestones and carbonic acids with rain waters into the forested hills activated the denudation process in the inner parts of the island. The steep slopes on the scarps are ranging from 23° to 55°, pedimented slopes on the hill fronts are varying from 15° to 22° and gentle break of slopes on the terraced banks varies from 8° to 14° in the island landscapes.

Geologically, the island is formed with the groups of Ophiolite suite, archipelago formation and recent to sub-recent formation dominated with Maffic Ultramaffics, chalk and white clay stones, Melville limestones, raised beach materials, coralline sands, coral rags and mangrove swamp clays. Little Andaman Island is fringed with the Bay of Bengal on the western shores and washed by Andaman Sea on the eastern shores. Thus, the island is affected by west coast swell waves and east coast swell waves from both sides of the oceanic environment. The island is elevated up to 100 mts in height in the south-central hills and they are encircled with a significant highland surface of 60–70 mts in heights. Scarpland surface is also visible on the rims of above-mentioned landforms. Other areas are dominated by reef terraces and ria-type shores (Fig. 2.5).

Physiographically, the island of South Andaman is categorized into eight major landform units. The landform diversity is reflected by assemblages of hill ridges

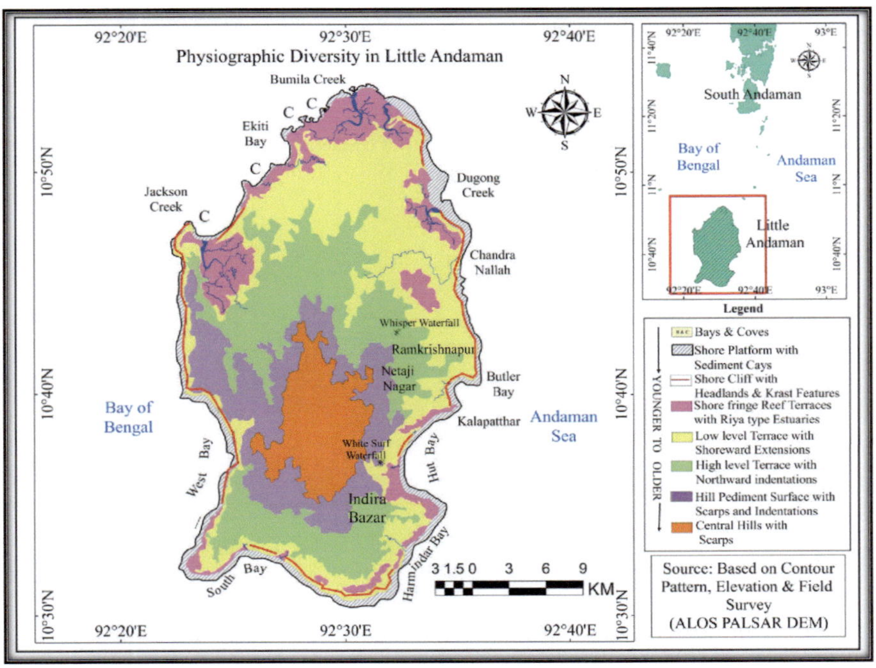

Fig. 2.5 Different landscape units of Little Andaman

with limestones; pediment slopes and valley flats; shore fringe reef terraces; inland terraces and cliffs of massive limestone terrain; ria-type estuaries and submerged valley floors; headlands with karst features and wave cut platforms; bays, coves and beaches along the seashores; and shore platforms with limestone pavements. The island is drained by radial drainage channels which are draining out from the central highland surface, and most of the drainage channels have produced several knick points and waterfalls along their longitudinal courses nearby the scarpland surfaces.

The 'Whisper Waterfall' and 'White surf Waterfall' with significant cliffs of limestone formation across the stepped surfaces appear as identical knick points in the South Central massif in Little Andaman. Most of the drainage channels along the shorelines of Harminder Bay, Hut Bay and Butler Bay were affected by raised shore features during the events of the earthquake (9.5 Richter scale) and tsunamigenic sediment deposits (2004, December 26).

The mouths of the east-flowing drainage channels were raised and blocked by sediments in the events of seismo-tectonic emergence and sediment transport. Over a long period after the tsunami incidence (2004), the inner sections of the drainage channels were ponded with water bodies and they have been converted into back waters at present (Fig. 2.6).

The shore platforms are highly wave-abraded, and joined with pools of water bodies and exposures of limestone pavements and apron of rubbles and sands. The

Fig. 2.6 Shore platform surface morphometry represented by clifflets, coral rags, carbonate sands and cavernous sediments, and Dugong creek mouth

headlands and shore cliffs in Kalapaththar section represent matures karst landforms under solution process (Fig. 2.4). Various caves, tunnels, sink holes, pinnacles, doline depressions and cavernous sediments are well developed in the joined limestones of Kalapaththar in between the Hut Bay and Buttler Bay sections of the eastern coast. Headlands and shore cliffs are notched by wave abrasion during the high tide levels along the shorelines of Little Andaman. Raised reef terraces of 5–10 m elevations are well preserved in the Harminder Bay and Hut Bay coasts. Over 96% areas of the island are densely covered by evergreen forests (Fig. 2.6).

Ria-type shores on the northern stretch are sediment filled and raised, the tidal flats frequent spilling tides and finer sediments provide ideal wetlands for colony of mangrove forests. All the creeks and east or northeast flowing channels are fringed with dense mangrove forests at their downstream sections in the island. Littoral forests and monsoon forests are located on the shore fringe dunes and reef terraces. In many places of the eastern coast, the intertidal mangrove wetlands have been raised by seismo-tectonic activities during 2004 (events of earthquake and tsunami) and converted into supratidal mangrove colonies. The landform units and their characteristics are illustrated in the following tables (Table 2.1).

2.1.2 Terrain Units (Cross-sectional Forms of Topography)

The island-wise terrain units have been represented with their physiographic characters, processes involved in the evolution of assemblages of features and diversity. Contour patterns helped to classify and to identify the terrain units of the island landscapes. Several cross-sections are drawn over the ALOS PALSAR DEM of each and individual island to highlight the topographic forms. The rate of weathering, denudation and solution process under the tropical moist climate on the local geologic structures, forested island platforms and bio-encrusted shore platforms have influenced in the modification of topographic forms (Figs. 2.7 and 2.8).

Highland and ridges have been denuded into residues, reef terraces, and intermontane valley flats have been raised, shore platforms and shore cliffs are modified into dynamic land units, drainage features have been altered into relict forms and sediments are accumulated into the beaches, coves and embayed shores of the islands. Topographic forms are well reflected in the island cross sections of the present study areas (Fig. 2.9).

2.1.3 Composition of Carbonate Platforms and Islands

Sediment composition of the shore platforms, beaches, sediment cays, beach rocks, terraces, valley flats and hill slopes are studied with their grain size analysis and material composition tests. The study reveals that seaward and landward parts of

Table 2.1 Landscape units and processes involved in Little Andaman

Little Andaman Island (landform diversities)						
Sl. no	Landform units	Elevation	Relief	Slope amount	Processes	Features
1	Hill ridges with limestones	60–100 m	40 m	16°–52°	Pedimentation fluviokarstic	Ridges, valleys, knick points, intervened valley flats, pediment surface
2	Pediment slopes and valley flats	60–80 m	20 m	9°–16°	Pedimentation, surface runoff	Terraced surface, pediment slopes, scarp with indentations
3	Shore fringe reef terraces	30–10 m	20 m	3°–6°	Subaerial weathering	Terraces, ria-type estuaries, mangrove wetlands
4	Inland terraces and cliffs of massive limestones	40–30 m	10 m	6°–9°	Surface weathering, solution process surface runoff	Scarps, waterfalls, knick points, sloping flats
5	Ria-type estuaries and submerged valley floors	6–10 m	04 m	3°–4°	Tidal and coastal marine	Estuaries, valley fill sediments
6	Headlands with karst features and wave cut platforms	10–25 m	15 m	3°–6°	Karstic process, wave abrasion	Caves, tunnels, solution pits, pinacles tafoni, dolines
7	Bays, coves, beaches along the shores	3–5 m	02 m	2°–2.85°	Waves, tides, currents, winds	Beaches, bars, caves, bays
8	Shore platforms with limestone pavements	4–8 m	04 m	2°–2.5°	Weathering. Wave abrasion	Wave cut platforms, limestone pavements

the shore platforms contain multiple sizes of grains as boulders, cobbles, pebbles, rubbles, gravels on the windward sides (Table 2.2, Fig. 2.10).

The shore fringe carbonate platforms are temporal sediment sink in which coral rubbles, debris, coral sands and other bioclasts are accumulated by waves and tides from the island fringe coral reefs. They are reworked and retransported along and across the shores during the events of large-scale marine inundations (storms,

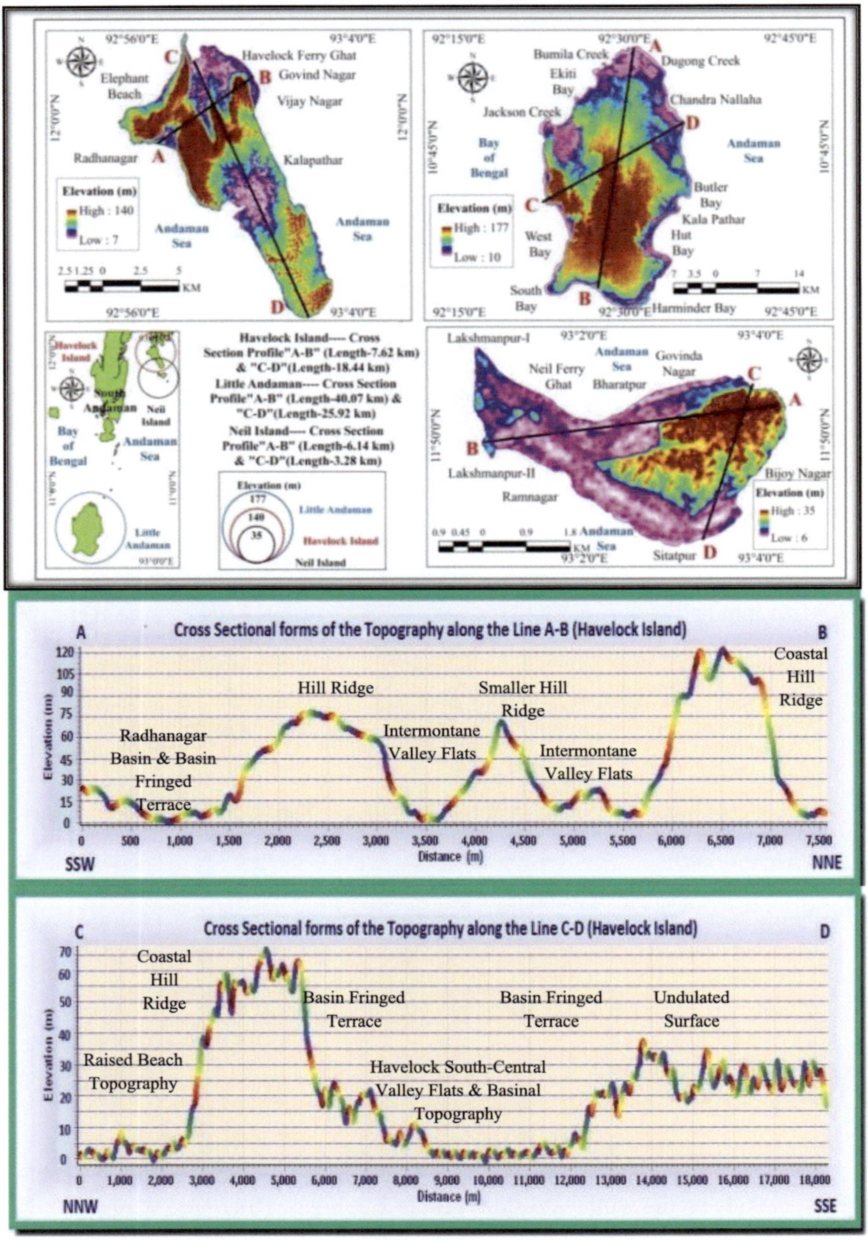

Fig. 2.7 ALOS PALSAR DEMS with topographic cross-sections (Havelock, Neil and Little Andaman)

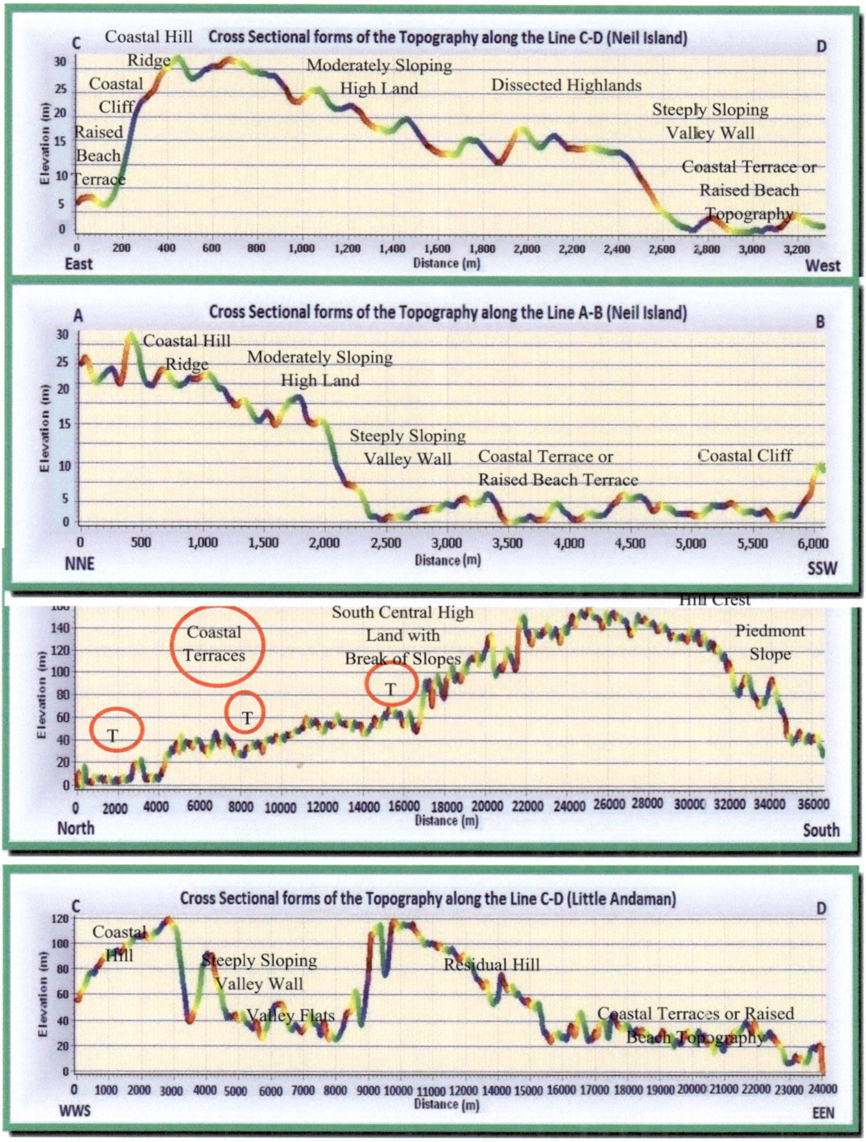

Fig. 2.8 Topographic cross-sections of Neil Island and Little Andaman

Fig. 2.9 Exposure of beach rocks, remnant of eroded reef terrace, promontory and fissure caves on the headland of the extended limestone rock body

Table 2.2 Grain size analysis of shore platform sediments in Havelock Island

Distribution of grain size in Havelock Island									
Name of beach	< 2000 μm	2000 μm	1000 μm	850 μm	500 μm	250 μm	125 μm	63 μm	Total weight
Radhanagar UB	0	113	0.062	0.88	256	0	0.008	0.05	370
Radhanagar LB	0	0	0	0	265	93	24	0	382
Kalapathar entrance UB	0	0.01	0.074	85.916	314	140	0	0	540
Kalapathar entrance LB	134	80	0.094	119.8	0.086	0.02	0	0	334
Kalapathar rock section UB	0.054	0.01	242	4.936	65	0	0	90	402
Kalapathar rock section LB	0	0.08	230	0	112	79.92	0	0	422

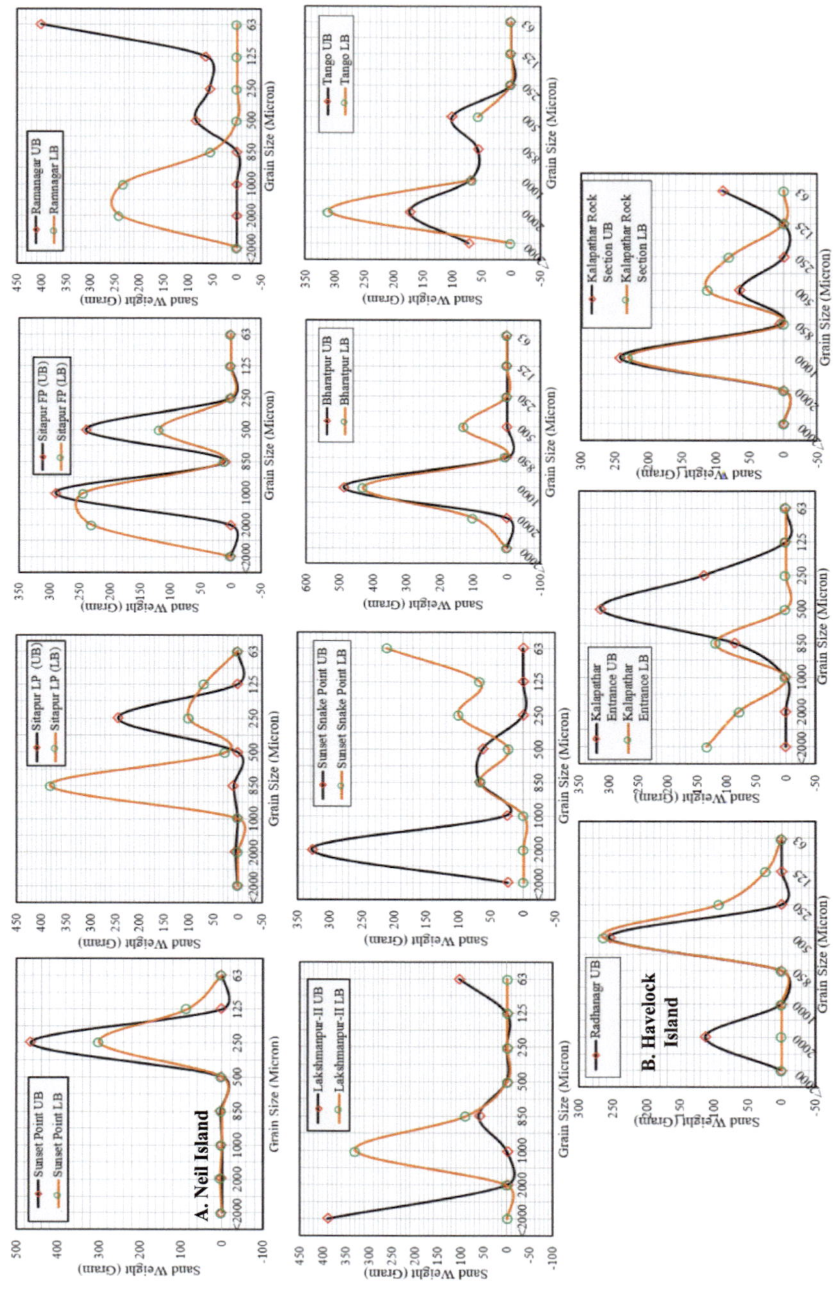

Fig. 2.10 Frequency curves showing the concentration of specific grain size in the shore platform of Neil Island and Havelock Island

tsunamis, swell waves) and finally accumulated in unconsolidated forms in a suitable position of the shore platforms. Such accumulation depends on the shore platform gradients and favourable shoreline configurations of the islands. Mass wasting, weathering process and stream deposition also supply sediments into the shore platforms from the landward side of the islands. The chemogenic sediments (*aragonites, calcites, calcium carbonates and dolomites)* also added with the other granular sediments of the shore platforms during tidal inundations twice daily in the tropical maritime zones. During immersion of tides when evaporation become strong, the unconsolidated sediments get cemented gradually over a time along the upper and middle parts of shore platform High Tide Line (HTL) into beach rocks. These beach rocks and bio-encrusted materials with the strong wave abrasive energies and weathering processes produce granular sediments into the shore platforms of the islands. Thus, material composition of the carbonate platforms is studied in this section by representing the grain size distribution graphs plotted particularly for the Havelock and Neil Islands of South Andaman region. The beach sands are characterized by 500 to over 1000-μm grain sizes of sediments, and the lower parts of the shore platform represent larger grain sizes dominated by 4000–5000 μm (Tables 2.2 and 2.3).

Table 2.3 Grain size analysis of shore platform sediments in Neil Island

Distribution of grain size in Neil Island									
Name of beach	< 2000 μm	2000 μm	1000 μm	850 μm	500 μm	250 μm	125 μm	63 μm	Total weight
Sunset point UB	0	0	0	4	0	466	0	0	470
Sunset point LB	0	4	0	0	0	300	84	0	388
Sitapur LP UB	0	4	0	8	0	244	0	0	256
Sitapur LP LB	0	0.68	0	382	24	100	67.32	0	574
Sitapur FP UB	0	0.012	290	9.988	240	0	0	0	540
Sitapur FP LB	0	230	244	11.988	118	0	0	0.012	604
Ramnagar UB	0	0.01	0.044	0	84	55.946	64	400	604
Ramnagar LB	0	240	232	53.946	0.046	0	0	0.008	526
Lakshmanpur-II LB	0.044	1.908	330	90	0.048	0	0	0	422
Lakshmanpur-II UB	390	0	0.13	60.836	0.034	0	0	103	554
Sunset snake point UB	24	328	27	68	63	0	0	0	510
Sunset snake point LB	0	0	0	67	23	100	68	212	470
Bharatpur LB	0	102	430	6	132	0	0	0	670
Bharatpur UB	0	2	488	0	0	0	0	0	490
Tango UB	72	172	69	57	100	0	0	0	470
Tango LB	0	310	66		56	0	0	0	432

The unconsolidated sediment grains are mostly overlying the lithified and cemented shore platforms, and the moistures surfaces are attached with brown and green algae mats. Such algal encrustations favour high rate of calcification in the tidal inundations and high rate of evaporation in the low tide exposures, which in turn they are developed into phytokarst features. Wave abrasive forces have generated shingle size grains of carbonate materials with skeletons of shells and algal grains on the shore platforms. Above the high tide level on the landward part of shore platforms, there is a continuous stretch of sand beach or gravel beach in the islands. Coralline sands carbonate sands, other bioclasts and coral rubbles are accumulated in the linear band of shore fringe platforms.

The shore cliff zone supplied materials by mass wasting process along the basal slope and fragmented into boulders and cobbles by wave energies during the events of large-scale marine inundations. Cliff erosion by wave actions also supplied sand size materials into the shore platform. Bodies of the rock boring animals, oysters, gastropods, marine shells and algal debris are thinly deposited over the extensive shore platforms and mixed with coral debris as organic sediments. They are lithified and cemented in favourable areas of high tide limits with the outflows of fresh water seepage lines and degassing of bicarbonates into the upper part of shore platforms.

These materials are known as beach rocks and they are composed of calcareous sands, silica sands, skeletons of corals and other bioclasts. They are cemented and lithified with aragonites, calcites, dolomites, calcium carbonates and oxides in the region. The stratified layers of calcareous sandstones, conglomerates, grits, shell limestones, clay stones and massive limestones also supply sediments into the base of shore cliff by wave erosion (Fig. 2.11).

The raised beach terraces of island platforms in Havelock Island and Neil Island are composed of shingles, sands, shells, stony corals and coral rags. Surface runoff on the pedimented hill slopes deposited carbonate clays, limestone debris and coral rags. The valley flats of the intermontane areas are composed of alluvium materials with clayey and loamy soils in the island interiors (Fig. 2.12).

Fig. 2.11 Coral rubbles, coralline sands, shore fringe sediments and the raise beach terrace sediment

Fig. 2.12 Measuring the coral rubbles, coralline sands, shore fringe sediments and the raise beach terrace sediment

2.1.4 Coastal Cliffs (Cliff Recession Rate)

The island fringe shore platforms in South Andaman are backed by narrow band of sandy beach and vertical wall of shore cliff. Shore cliffs are relatively higher on the headlands, coastal hills, and promontories. Moderately high cliffs or the medium height cliffs are visible along the bank margins of marine terraces or the raised beach platforms in the islands. Low cliffs are usually found along the sediment cays and beach fringed banks of the carbonate islands.

The structure of the sea cliffs is categorized into horizontally stratified layers of carbonate rocks, massive limestones, vertical wall rocks with white clay stones or chalk, landward tilted bedding planes of alternate rock layers dominated with laminated shelly limestones, white clay stones, calcareous sandstones and boulder conglomerates with debris of corals and fossiliferous deposits (Fig. 2.13). Shore cliffs on reef terraces are composed of reefal rocks with coralline debris, marine shells, and other bioclasts. Alternate hard rocks and soft rocks, compactness or strength of the rocks, granular rocks with coarser sands, grits, shingles and conglomerates, jointed and fractured or faulted rocks and rocks of finer grains influenced the erosion rates of shore cliffs in the oceanic environment.

Hydrologically, the shore cliffs may be affected by seepage pits, percolation of fresh waters through the joints, root-penetrated zones with drips of infiltrated waters and flank margin water table leakages through cliff walls during the period of monsoon rains in the islands. Weathering and mass wasting are enhanced by the hydrological effects over the carbonated soluble rocks on the shore cliffs (Fig. 2.14). Large-scale events of marine inundations with significant wave heights undermined or notched the base of cliff walls in many places of the islands which allowed the vertical cracked wall rocks to collapse. Seismo-tectonic activities also increased stress within the jointed rocks of the shore cliffs to develop processes of debris fall in the shore platforms in the region of South Andaman Islands. Active weathering also promoted on the cliff walls by wave splashing and wave breakers in the monsoon season particularly along the sea face of windward sides.

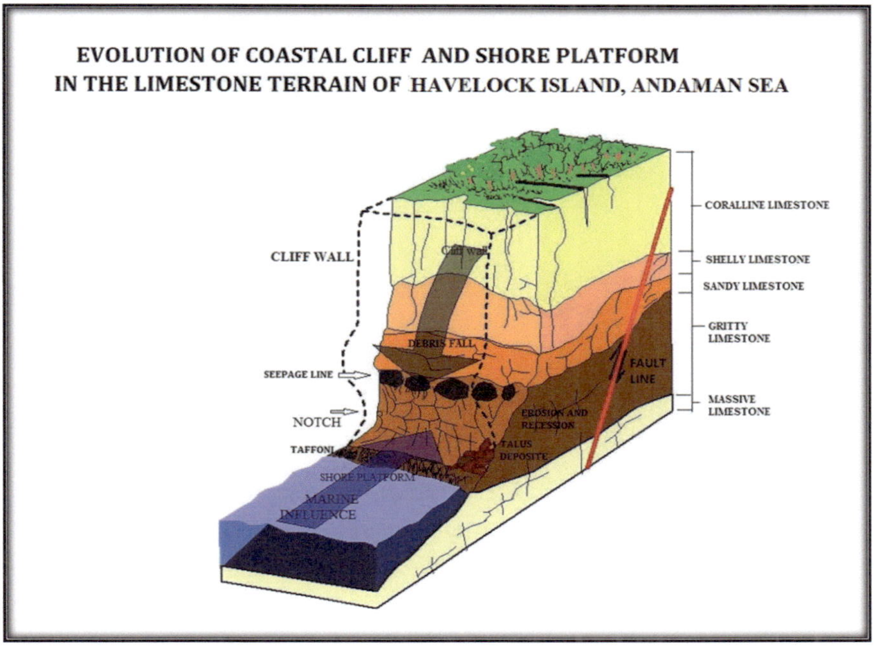

Fig. 2.13 Schematic diagram showing the reasons for cliff recession

All the above factors are responsible for cliff failures and their recession or erosion in the carbonate islands of South Andaman. Temporal remote sensing data (Table 2.4, Figs. 2.15 and 2.16) is used to identify and delineate the cliff lines of different years along the island shores. The study shows that significant temporal changes are recorded as 0.28–0.48 m/yr recession rate for Havelock Island, 0.24–0.26 m/yr for Neil Island and 0.16–0.66 m/yr recession rate for Little Andaman. The seaward edges of shore platforms on the other hand are buttressed due to strong wave actions particularly on the windward reef fronts. Steeply sloping reef front bank is also advancing landward rapidly at a rate of wave erosion in the island fringe shore platforms. It is estimated that there is an equilibrium condition between the erosion rates of shore cliffs and reef front banks in few places (Fig. 2.14).

However, differences of recession rate between cliff lines and shore platform bank lines in places indicated near equilibrium condition due to local geology and windward reef front location. The existence of shore platforms backed by shore cliffs exhibited such equilibrium condition in the islands of South Andaman. Shore platforms width is reduced in which the erosion rate is rapid (Tables 2.5, 2.6 and 2.7; Fig. 2.17).

Geologically, the coastal cliffs of Little Andaman, Havelock and Neil Island appeared as horizontal layers, tilted planner shaped layers of sedimentary rocks and vertical to sub-vertical rocks, and fronted by low-gradient shore platforms. They are

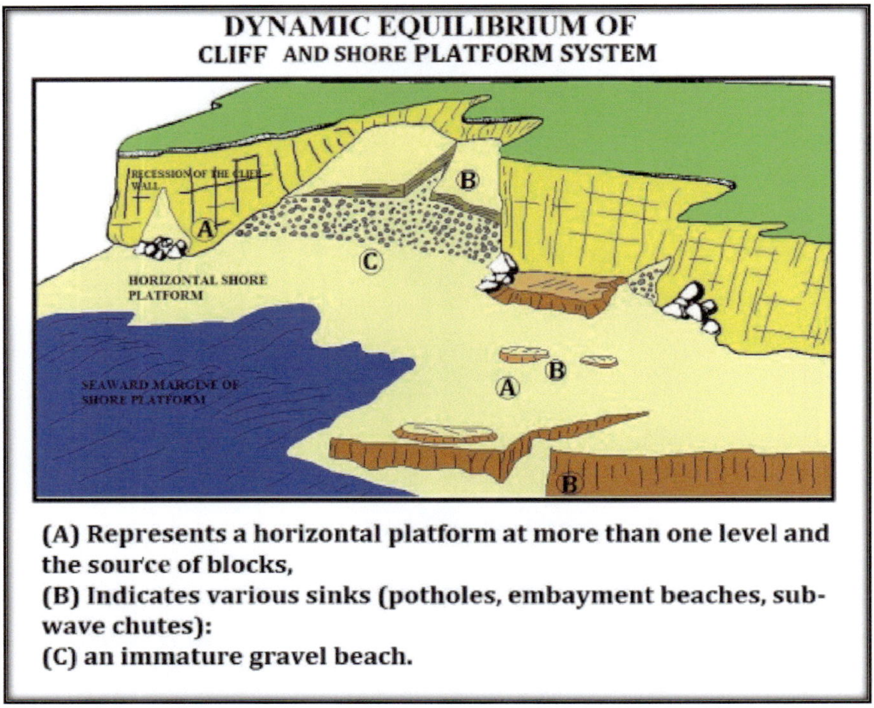

Fig. 2.14 Equilibrium condition between the recession rate of the shore cliff and bank line cliff

Table 2.4 Recession rate of the shore cliff and shore platform banks in carbonate islands (Havelock, Neil and Little Andaman)

Island	Shore cliff recession (1993–2018)		Shore platform bank line recession (1993–2018)	
	Total in m	Rate per year	Total in m	Rate per year
Havelock Island	6.46–14.47 m/25 yr	0.28–0.48 m/yr	13.98 m–22.86 m/25 yr 12.18 m–44.22 m/25 yr	0.93 m–1.52 m/y 0.81 m–2.95 m/y
Neil Island	5.54 m–6.65 m/25 yr	0.24 m/yr–0.26 m/yr	56.28 m–19.44 m/25 yr 9.51 m–8.23 m/25 yr	0.78 m–2.26 m/y 0.33 m–0.39 m/y
Little Andaman	3.98 m–16.41 m/25 yr	0.16 m/yr–0.66 m/yr	45.97 m–10.94 m/25 yr 32.39 m–40.92 m/25 yr	1.99 m–2.12 m/y 1.29 m–1.64 m/y

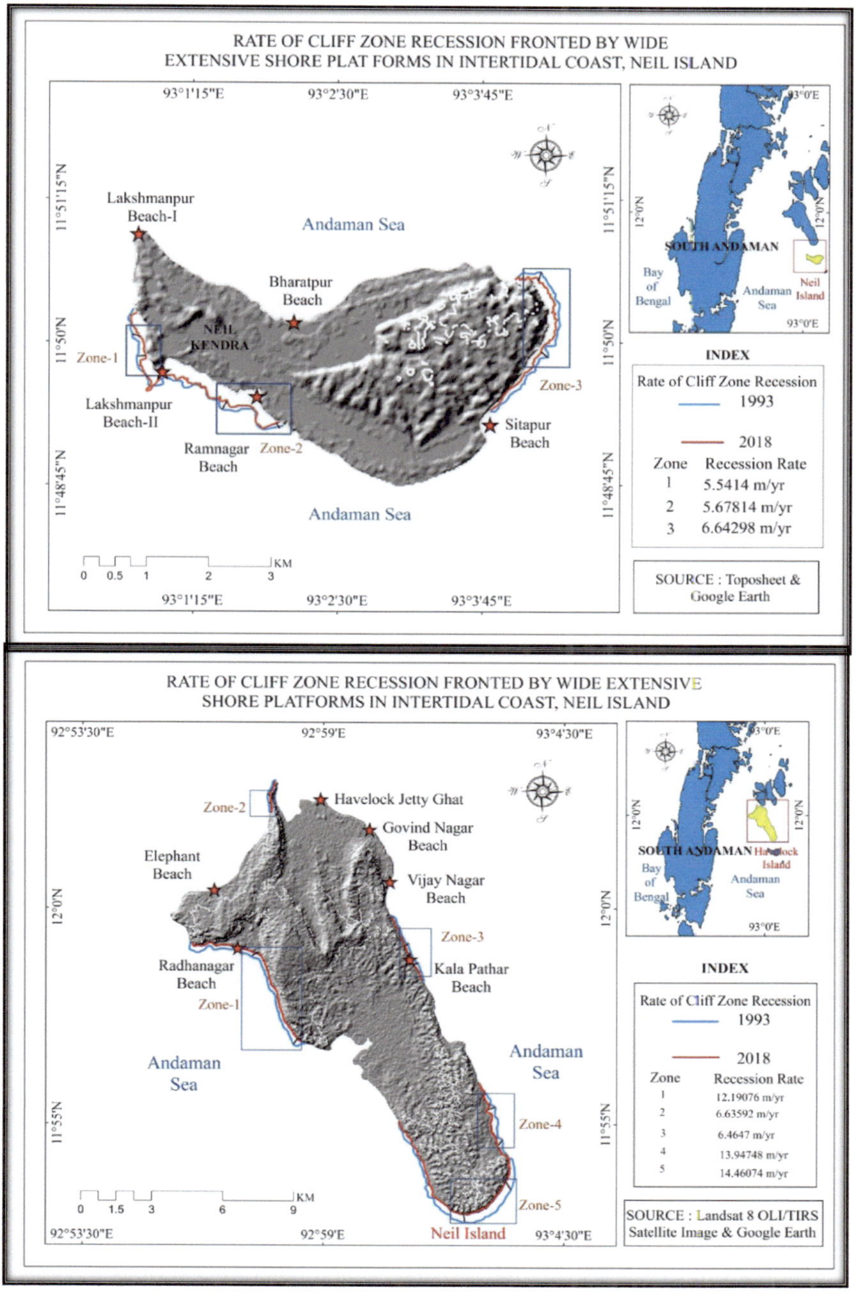

Fig. 2.15 Assessment of the cliff recession diversity in the shoreline sections of Neil Island and Havelock Island using SAR data (C band tata)

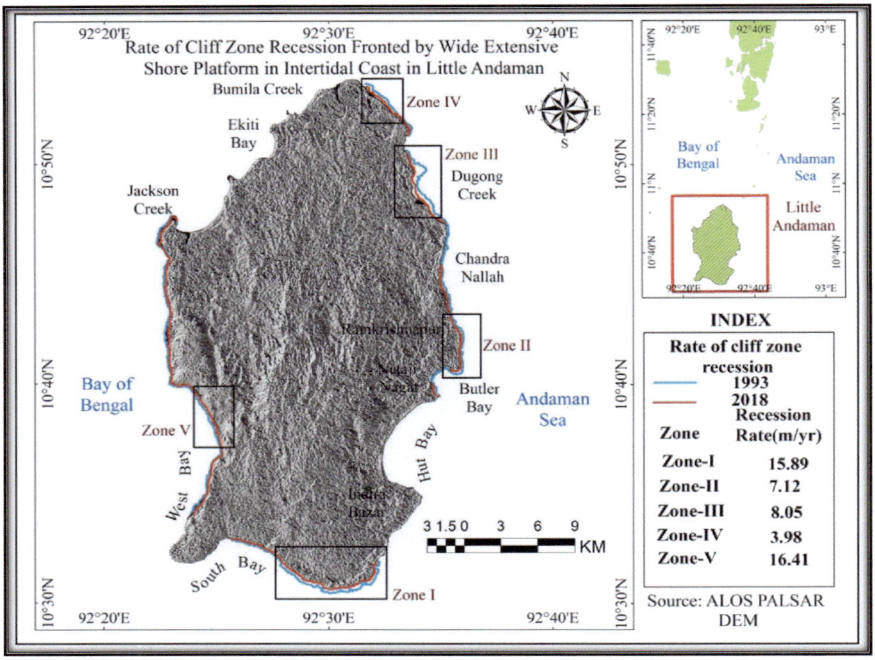

Fig. 2.16 Estimation of cliff recession rate in Little Andaman using SAR data

jointed, fractured and faulted in different sections. The geology, lithology, hydrology, tectonics, weathering rate, mass wasting, sea level, events of marine inundations and coastal hydraulics act as the process dynamics involved in the recession of shore cliffs in the tropical climate. The carbonate rocks of limestone, gritty sandstones, white clay stones, marls or chalky limestones, shelly limestones and coralline limestones of the shore cliffs are also soluble under the water seeps areas in the island coast. Wave splashing and tidal current scours displayed open marine energy concentration at the base of the shore cliffs. Thus, the shore cliff recession rate differs from place to place in the region.

The shore platform width depends on recession rate of the cliff wall. On the windward side of the island, shore cliffs are fronted by narrow shore platform surface particularly due to the rapid erosion of platform edge on the sea face. Probably, there is a strong relationship between the shore cliff recession and shore platform margin bank erosion rate in the region.

The shore cliff recession rates of the islands are compared with temporal scale and spatial scale changes. The satellite image data of 2005, 2010, 2012 and 2019 are overlaid in a similar scale to get the active recession rate of the various sections of shore cliffs in the region.

However, field validation is conducted with monitoring changes of the shore cliff sections of the islands. The study reveals that headlands and promontories are mostly

Table 2.5 Database on cliff recession rate tabulated for Neil Island

Rate of Cliff zone recession of Neil Island (1993–2018)				
Location	Latitude	Longitude	Elevation (m)	Cliff rate recession 1993–2018 (m/y)
Lakshmanpur-I (Z-1)	11°50′ 24.83″ N	93° 0′ 45.83″ E	9	5.12
	11°50′ 22.22″ N	93° 0′ 45.86″ E	10	2.39
	11°50′ 16.75″ N	93° 0′ 47.19″ E	13	4.52
	11°50′ 11.29″ N	93° 0′ 49.72″ E	12	3.02
	11°50′ 7.74″ N	93° 0′ 52.09″ E	13	5.55
	11°50′ 1.67″ N	93° 0′ 53.01″ E	14	4.69
	11°49′ 56.46″ N	93° 0′ 50.86″ E	12	4.02
	11°49′ 48.63″ N	93° 0′ 54.90″ E	6	1.06
	11°49′ 43.02″ N	93° 0′ 54.97″ E	2	3.02
Ramnagar Beach (Z-2)	11°49′ 48.36″ N	93° 0′ 58.73″ E	11	5.68
	11°49′ 51.07″ N	93° 1′ 2.24″ E	8	5.23
	11°49′ 51.63″ N	93° 1′ 7.73″ E	9	2.36
	11°49′ 51.06″ N	93° 1′ 12.74″ E	11	1.29
	11°49′ 48.04″ N	93° 1′ 18.92″ E	9	4.23
	11°49′ 44.48″ N	93° 1′ 26.20″ E	8	5.57
	11°49′ 42.21″ N	93° 1′ 29.18″ E	7	3.69
	11°49′ 41.65″ N	93° 1′ 32.42″ E	6	4.5
	11°49′ 38.82″ N	93° 1′ 37.89″ E	10	5.5
	11°49′ 37.21″ N	93° 1′ 45.29″ E	8	4.59
	11°49′ 29.21″ N	93° 1′ 52.31″ E	7	5.05
	11°49′ 23.25″ N	93° 1′ 54.59″ E	5	3.69
	11°49′ 19.67″ N	93° 1′ 59.33″ E	8	2.59
	11°49′ 16.92″ N	93° 2′ 6.58″ E	13	3.98
Sitapur Beach (Z-3)	11°49′ 24.62″ N	93° 3′ 45.99″ E	27	5.98
	11°49′ 33.22″ N	93° 3′ 53.78″ E	24	6.53
	11°49′ 41.01″ N	93° 4′ 1.31″ E	28	2.9
	11°49′ 50.30″ N	93° 4′ 7.26″ E	23	4.03
	11°49′ 56.12″ N	93° 4′ 10.89″ E	32	3.12
	11°50′ 1.19″ N	93° 4′ 15.43″ E	19	4.06
	11°50′ 7.93″ N	93° 4′ 16.27″ E	27	2.69
	11°50′ 12.06″ N	93° 4′ 15.26″ E	38	3.89
	11°50′ 17.46″ N	93° 4′ 15.70″ E	33	6.65
	11°50′ 21.49″ N	93° 4′ 13.39″ E	49	4.03
	11°50′ 26.20″ N	93° 4′ 10.73″ E	39	5.09
	11°50′ 29.60″ N	93° 4′ 6.36″ E	18	5.29

(continued)

Table 2.5 (continued)

Rate of Cliff zone recession of Neil Island (1993–2018)				
Location	Latitude	Longitude	Elevation (m)	Cliff rate recession 1993–2018 (m/y)
	11°50′ 29.22″ N	93° 4′ 0.95″ E	34	1.29
	11°50′ 29.84″ N	93° 3′ 54.91″ E	31	2.19
	11°50′ 34.46″ N	93° 3′ 50.24″ E	14	3.12
	11°50′ 38.24″ N	93° 3′ 45.06″ E	10	6.29

Table 2.6 Database on cliff recession rate tabulated for Havelock Island

Rate of cliff zone recession of Havelock Island (1993–2018)				
Location	Latitude	Longitude	Elevation (m)	Cliff rate recession 1993–2018 (m/y)
Radhanagar (Zone-1)	11°59′ 8.60″ N	92°56′ 7.30″ E	22	8.51
	11°59′ 7.97″ N	92°56′ 16.92″ E	18	6.02
	11°59′ 10.99″ N	92°56′ 22.48″ E	20	5.39
	11°59′ 8.83″ N	92°56′ 28.61″ E	21	5.06
	11°59′ 7.66″ N	92°56′ 38.35″ E	19	9.25
	11°59′ 7.61″ N	92°56′ 52.16″ E	20	5.62
	11°59′ 4.81″ N	92°57′ 2.04″ E	15	7.09
	11°59′ 3.44″ N	92°57′ 18.35″ E	11	12.19
	11°58′ 54.69″ N	92°57′ 31.86″ E	13	8.94
	11°58′ 36.81″ N	92°57′ 42.85″ E	38	9.52
	11°58′ 16.37″ N	92°57′ 48.57″ E	23	4.89
	11°57′ 56.93″ N	92°57′ 57.90″ E	25	6.01
	11°57′ 40.64″ N	92°58′ 0.94″ E	18	11.95
	11°57′ 24.25″ N	92°58′ 6.42″ E	15	8.05
	11°57′ 11.05″ N	92°58′ 15.12″ E	25	10.29
	11°56′ 56.63″ N	92°58′ 24.81″ E	26	3.29
	11°56′ 51.21″ N	92°58′ 38.94″ E	27	10.75
Upper part of Elephant Beach (Z-2)	12° 2′ 14.27″ N	92°57′ 49.02″ E	15	4.06
	12° 2′ 22.04″ N	92°57′ 49.06″ E	28	3.59
	12° 2′ 28.54″ N	92°57′ 47.98″ E	14	5.89
	12° 2′ 35.37″ N	92°57′ 46.84″ E	24	6.64
	12° 2′ 42.23″ N	92°57′ 48.21″ E	33	4.79
	12° 2′ 46.30″ N	92°57′ 50.91″ E	25	3.09
	12° 2′ 52.75″ N	92°57′ 52.83″ E	17	5.29

(continued)

Table 2.6 (continued)

Rate of cliff zone recession of Havelock Island (1993–2018)				
Location	Latitude	Longitude	Elevation (m)	Cliff rate recession 1993–2018 (m/y)
Kalapathar Beach (Z-3)	11°59′ 36.84″ N	93° 0′ 34.95″ E	19	6.99
	11°59′ 23.20″ N	93° 0′ 40.79″ E	27	5.98
	11°59′ 7.84″ N	93° 0′ 48.99″ E	27	2.35
	11°58′ 56.57″ N	93° 0′ 54.21″ E	31	4.5
	11°58′ 46.87″ N	93° 0′ 57.78″ E	25	6.45
	11°58′ 34.52″ N	93° 1′ 5.85″ E	21	4.59
	11°58′ 22.98″ N	93° 1′ 11.35″ E	18	6.01
	11°58′ 8.52″ N	93° 1′ 19.41″ E	20	4.85
	11°57′ 58.25″ N	93° 1′ 25.97″ E	23	3.78
	11°57′ 48.42″ N	93° 1′ 29.71″ E	29	5.96
Lower part of Kalapathar Beach (Z-4)	11°55′ 49.26″ N	93° 2′ 34.10″ E	20	8.06
	11°55′ 32.48″ N	93° 2′ 38.12″ E	23	6.75
	11°55′ 23.00″ N	93° 2′ 41.93″ E	12	12.11
	11°55′ 11.66″ N	93° 2′ 44.10″ E	14	9.09
	11°54′ 55.49″ N	93° 2′ 46.10″ E	19	10.59
	11°54′ 37.08″ N	93° 2′ 53.41″ E	14	13.95
	11°54′ 20.59″ N	93° 3′ 0.53″ E	16	11.12
Lower part of Havelock Island(Z-5)	11°54′ 4.07″ N	93° 3′ 12.20″ E	38	5.09
	11°53′ 51.40″ N	93° 3′ 11.63″ E	40	11.09
	11°53′ 37.63″ N	93° 3′ 7.26″ E	34	9.06
	11°53′ 25.84″ N	93° 3′ 0.58″ E	30	8.75
	11°53′ 17.45″ N	93° 2′ 53.66″ E	39	10.59
	11°53′ 10.07″ N	93° 2′ 47.26″ E	36	6.23
	11°53′ 0.42″ N	93° 2′ 34.97″ E	22	4.36
	11°52′ 55.10″ N	93° 2′ 20.21″ E	38	14.47
	11°52′ 56.69″ N	93° 2′ 6.19″ E	25	11.02
	11°53′ 5.10″ N	93° 1′ 50.27″ E	39	7.09
	11°53′ 14.79″ N	93° 1′ 37.30″ E	14	10.9
	11°53′ 30.50″ N	93° 1′ 31.11″ E	18	9.05
	11°53′ 44.01″ N	93° 1′ 26.86″ E	15	12.48
	11°53′ 58.54″ N	93° 1′ 22.79″ E	26	5.5
	11°54′ 10.41″ N	93° 1′ 15.70″ E	25	7.72
	11°54′ 29.93″ N	93° 1′ 7.03″ E	21	8.06
	11°54′ 46.51″ N	93° 0′ 56.39″ E	30	6.09
	11°54′ 56.40″ N	93° 0′ 47.66″ E	24	13.59

Table 2.7 Database on cliff recession rate tabulated for Little Andaman

Rate of cliff zone recession of Little Andaman (1993–2018)				
Location	Latitude	Longitude	Elevation (m)	Cliff rate recession 1993–2018 (m/y)
South Bay (Z-1)	10°32′ 59.29″ N	92°25′ 51.32″ E	12	11.09
	10°32′ 47.06″ N	92°26′ 31.13″ E	11	10.29
	10°32′ 30.49″ N	92°27′ 14.71″ E	24	13.29
	10°32′ 3.06″ N	92°27′ 53.73″ E	30	9.74
	10°31′ 18.27″ N	92°28′ 40.96″ E	23	11.09
	10°31′ 2.41″ N	92°29′ 21.50″ E	13	15.91
	10°30′ 41.47″ N	92°30′ 7.22″ E	17	8.36
	10°30′ 59.07″ N	92°31′ 19.27″ E	25	14.29
	10°31′ 20.11″ N	92°31′ 57.54″ E	20	12.09
	10°32′ 5.02″ N	92°32′ 12.89″ E	24	14.29
Butler Bay (Z-II)	10°39′ 12.93″ N	92°34′ 18.10″ E	21	7.12
	10°39′ 42.67″ N	92°34′ 31.28″ E	14	6.59
	10°40′ 23.94″ N	92°34′ 31.20″ E	26	5.09
	10°40′ 50.07″ N	92°35′ 12.36″ E	21	7.01
	10°40′ 49.40″ N	92°35′ 45.53″ E	20	6.66
	10°41′ 32.83″ N	92°35′ 41.05″ E	27	3.09
	10°42′ 25.65″ N	92°35′ 35.78″ E	22	4.5
	10°43′ 17.03″ N	92°35′ 16.55″ E	23	6.99
Dugong Creek (ZIII)	10°45′ 3.57″ N	92°34′ 53.84″ E	26	7.56
	10°46′ 14.74″ N	92°35′ 3.03″ E	8	4.12
	10°47′ 19.02″ N	92°34′ 51.53″ E	22	6.59
	10°47′ 59.06″ N	92°34′ 9.28″ E	13	7.26
	10°48′ 41.89″ N	92°33′ 40.35″ E	11	8.05
	10°49′ 17.72″ N	92°33′ 33.22″ E	18	5.96
	10°49′ 47.46″ N	92°33′ 37.42″ E	9	6.99
	10°50′ 17.87″ N	92°33′ 26.10″ E	7	8.01
	10°50′ 55.19″ N	92°33′ 20.39″ E	12	7.26
Upper part of Dugong Creek (Z-IV)	10°51′ 27.04″ N	92°33′ 28.80″ E	10	2.59
	10°52′ 3.60″ N	92°33′ 1.96″ E	15	1.26
	10°52′ 27.68″ N	92°32′ 35.88″ E	12	2.56
	10°52′ 42.85″ N	92°32′ 20.36″ E	10	3.99
	10°53′ 0.15″ N	92°31′ 52.16″ E	4	3.9
	10°53′ 20.80″ N	92°31′ 37.33″ E	13	2.5
	10°53′ 36.75″ N	92°31′ 15.46″ E	2	2.79
West Bay (Z-V)	10°33′ 18.65″ N	92°23′ 39.95″ E	15	3.89

(continued)

Table 2.7 (continued)

Rate of cliff zone recession of Little Andaman (1993–2018)

Location	Latitude	Longitude	Elevation (m)	Cliff rate recession 1993–2018 (m/y)
	10°34′ 22.05″ N	92°24′ 7.65″ E	18	12.96
	10°35′ 17.08″ N	92°24′ 37.52″ E	21	10.23
	10°36′ 11.06″ N	92°24′ 49.13″ E	40	14.29
	10°37′ 28.55″ N	92°25′ 7.92″ E	36	15.23
	10°38′ 46.02″ N	92°24′ 36.00″ E	30	9.23
	10°39′ 39.87″ N	92°24′ 0.23″ E	22	13.79
	10°39′ 59.60″ N	92°23′ 4.36″ E	17	16.41
	10°41′ 7.62″ N	92°22′ 54.65″ E	14	15.96
	10°41′ 52.27″ N	92°22′ 50.43″ E	19	10.96
	10°43′ 19.01″ N	92°22′ 59.47″ E	16	11.29
	10°44′ 31.82″ N	92°22′ 44.79″ E	25	11.02
	10°45′ 35.15″ N	92°22′ 54.35″ E	22	12.78
	10°46′ 36.72″ N	92°22′ 34.79″ E	11	14.02

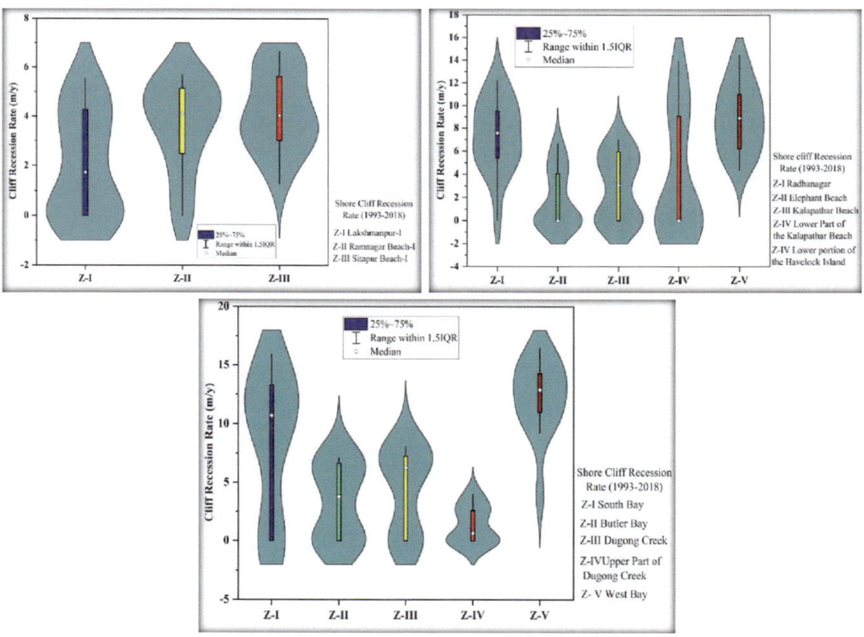

Fig. 2.17 Box plots showing the median and quartile on shore cliff recession rate for Neil Island, Havelock Island and Little Andaman

affected by shore cliff recession at the rate of 0.89 mts in between 2005 and 2019 (14 years of time span) in the region of Havelock Island.

2.1.5 Stratigraphy and Facies Analysis

Suitable exposure sections of stratigraphy are studied in three different islands for environmental facies analysis. Radhanagar west section of Havelock Island represents seven successive layers of sedimentary environment with 5.96 m sea cliff wall extended from the shore platform. The section is also affected by a normal fault in the region. The tilted bedding planes towards the land at an angle of 40°–45° dip amount with the strike value of N 65° W appeared as sea cliff in anti-dip slope along the shoreline of Radhanagar west (Fig. 2.18). A dense area of tropical evergreen forest is located on the crest of shore cliff and also on the landward slope of the tilted hill ridge in the region. The top layer of the shore cliff is composed of carbonate clays with marine shells and dense root zone of vegetations. The second layer of the cliff represents conglomeratic debris with fossil corals and decomposed coral rags and bi-valve shells. The third layer from the top appears as massive clay stone with calcareous deposits. However, the layer-4 on the shore cliff is constituted with thinly laminated clay stone, sandstone and clay stone sheets. Very compact and sheety clay stone occupies the layer-5 in the vertical sequence of the shore cliff. The sixth successive layer is made up with laminated sandstone and micro folding structures in a continuous manner. Massive and sheety sandstone in a discontinuous form is visible on the shore cliff below the sixth layer of sedimentary deposits. The bottom layer is relatively thicker with white clay stone material that behaves differently in massive and jointed forms on the sea face.

The lithological succession layers of sediments reveal that all the sedimentary materials are formed under the Fore Arc Basin deposits and Quaternary boulder deposits in a shallow marine environment of Eocene, Miocene and Pliocene periods. They were isostatically raised, tilted, and deformed with the local tectonics under a stress field of Island Arc settings. The calcareous carbonate deposits and coral reef building environment existed during Holocene transgressive phase along the shores of the region exposure section. This section was cracked, collapsed and subsided after the seismo-tectonic event of 2004 in this island (Fig. 2.19).

A lithological succession section is also studied in the Hut Bay near Kalapathhar in Little Andaman Island. The Headland section of Little Andaman represents a dark grey calcareous mudstone at the base and the thick bed is unconformably underlain by a depositional bed with shell limestone with coral rags and raised beach materials.

The calcareous mudstone with ash beds produced various pseudokarstic landforms by solution, weathering and wave erosion processes in the region. Entire formation of the section belongs to middle Miocene and Late Pleistocene ages. The depositional beds of shell limestone with coral rags are isostatically raised from the shallow marine coralline environment, and denuded at the top of the headlands by weathering and erosion during recent-sub-recent stages (Figs. 2.19 and 2.20).

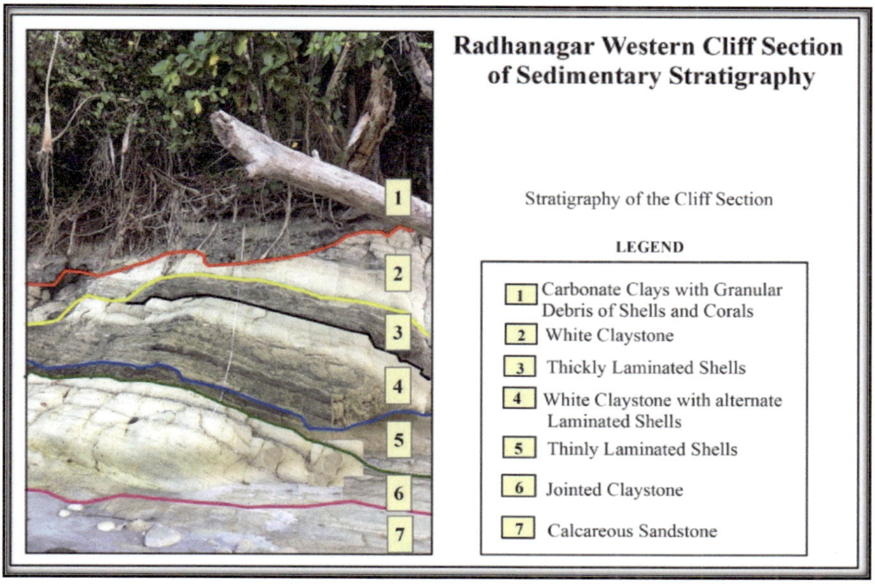

Fig. 2.18 Stratigraphic records on the field survey documented on Havelock Island

Fig. 2.19 Stratigraphic records on the field survey documented on Little Andaman (Kalapaththar section)

Fig. 2.20 Bio-calcarenites on the stratigraphic exposures on Neillian West formation

The lithological successions are studied in three exposure sites of Neil Island (Neil west, Lakshmanpur, and Sitapur) to explain their environmental facies (Figs. 2.21 and 2.22). There is a deformed cliff section of 6.5–7.2 mts altitude over the shore platforms on the western shore of Neil Island. The section comprises 5 layers of sedimentary beds dominated by limestone and silty mudstone members of geological formation. Former studies by Srinivasan and Azmi (1976a) indicated that the Neil West Coast formation of the Archipelago Group belongs to Early Pleistocene to Late Pliocene Age. The Neil limestone formation is unconformably lying over the silty mudstone member of Late Pliocene Age. The field study reveals that the stratigraphical layers from the top includes medium-grained algal bio-calcarenites with compact hard structures, sheety and laminated shell limestones of fine- to medium-grained materials, fossiliferous coralline limestones with coarse-grained cemented materials, conglomeratic coral rubbles with calcareous sandstones and shell fragments, compact and massive limestones of medium-grained to graveliferous materials of yellowish colour at the basement.

The average thickness of all the beds is ranging from 1.2 to 1.4 m in the sedimentary beds of the stratigraphic section represented ancient coastal, shallow marine and coral reef environments, with interactions of isostatic upliftment and sea level fluctuations of the regional geologic settings, and modern weathering processes of the tropics.

Another lithological succession of Neil Island was studied nearby the Natural Arch (Howrah Bridge) of Lakshmanpur-II in which 20 m exposure of different carbonated sedimentary beds are visible at the low-tide shore platforms. The silty mudstone layer is unconformably underlying the overlain beds of medium-grained materials. From the bottom of the section the second bed consists of shell limestone with boulders and gravels; calcareous sandstones with coarser grained materials are composed along the next bed of stratigraphy of the exposure site. The fourth consecutive bed from the bottom comprises sheety limestone with creamy colour compact and fine-grained materials and micro-folded structures in the stress field. The bed of massive limestone with gritty materials and coral rags occupies the next layer in stratigraphy. Exposure of the bed is blanketed by decomposed algal encrustation of calcareous materials. However, the top most bed of the section is composed of black muds of mangrove swamps in carbonated form, as youngest deposits, and exposed along the sedimentary layer (Fig. 2.22).

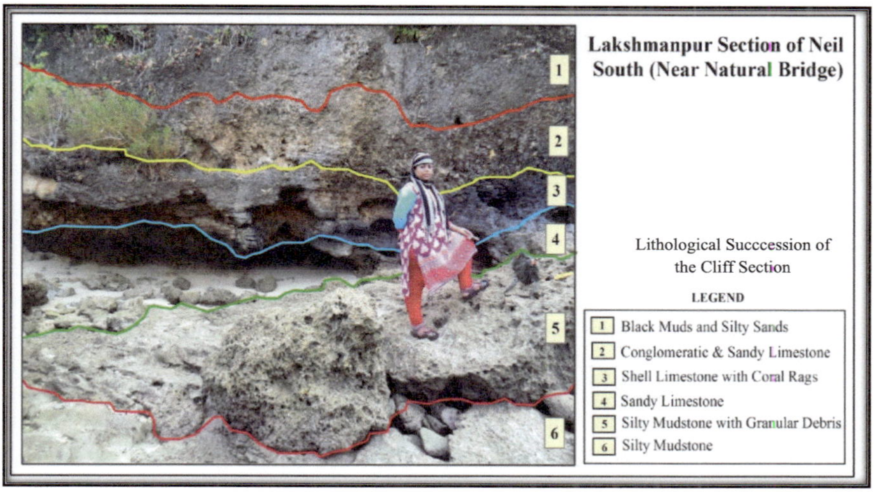

Fig. 2.21 Stratigraphic records documented from the Neil west section nearby the Natural Arch

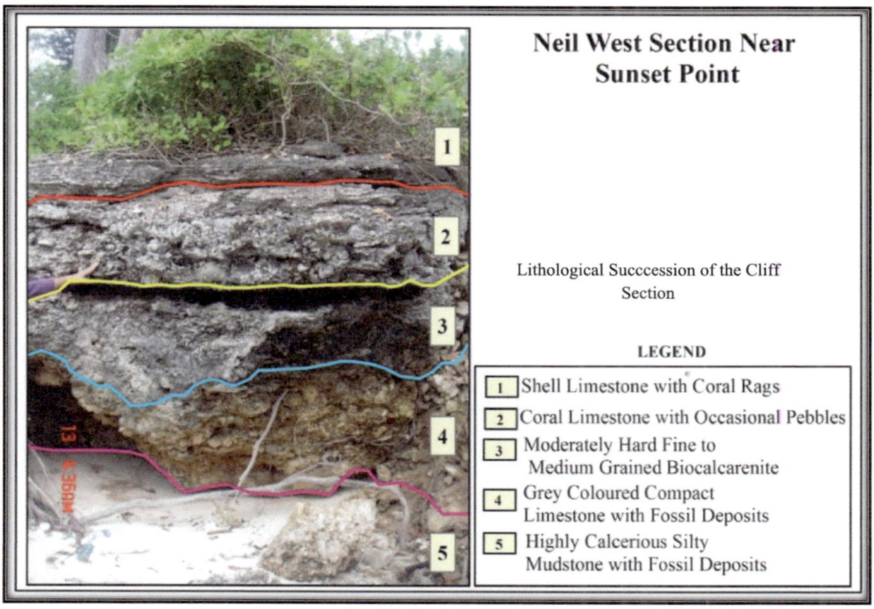

Fig. 2.22 Stratigraphic records documented from the Neil west section nearby the Sunset Point

In another site nearby the section of Neil West, the elevated bed of shell limestones with coral rags at an altitude of 15.5 m above the shore platform exhibits the indication of raised beach surface in the island fringe setup. The environmental facies of sedimentary beds represent depositional sequences of shallow marine, coastal carbonates with coral fringe environment, and intertidal mangrove environment in the region of Mio-Pliocene to Late Pleistocene Age. Recent upliftment took place after the formation of depositional sequences of calcareous sediments in the island (Fig. 2.23).

The Sitapur section of Neil East is exposed on the shore cliff areas of 20 mts altitude from the baseline of shore platform. There is an exposure of white clay stone bed at the base with compact and hard surface on the East Coast Section. The shell limestone and coral rags with thick deposits of medium-grained materials are lying over the underlying bed in the section. Towards northeast of the section the entire exposure is composed of soft, light to bluish grey, calcareous, massive mudstone with occasional thin siltstone bands and calcareous concretions. The massive mudstone of this section appears as brittle with development of sub-vertical joints, fractures and faulting under the stress field. Many fissure caves are formed at the base of this calcareous mudstone by solution and wave erosion processes in concentrated forms along the joints and fissures. The siltstone bands and calcareous concretions along the fault planes are crashed in the stratigraphic section. Cliff top failures with rock falls and earth materials as well as uprooted logs appeared as basal slope in depositional forms along the intertidal shore platforms of the exposure section (Fig. 2.23).

The lithological succession layers of Sitapur section exhibited the ancient depositional environment of deeper marine and shallow marine conditions of Late Miocene, Early Pliocene to Late Pleistocene epochs. Shell limestone with coral rags probably

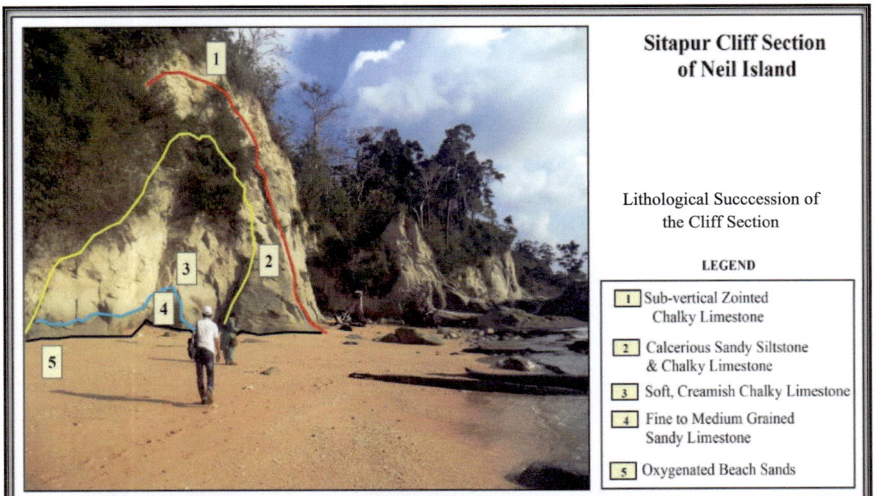

Fig. 2.23 Stratigraphic record documented in Sitapur cliff section of Neil Island

formed in a coralline environment. The slow upliftment of the beds took place in the Late Pleistocene epoch and gradually deformed with the tectonic events during Holocene age in the region. Sea level fluctuations, weathering and wave abrasion have resulted the current development of the cliff section exposures (Fig. 2.23).

2.1.6 Reef Terraces and Landforms

The shallow marine fringe coral banks along the island shores of Little Andaman, Havelock and Neil Island are raised successively as reef terraces in the Pre-Holocene age. Terraces from the seaward sides to the landward parts comprise age old materials of reefal rocks, limestones with coral rags, fossiliferous coral debris, sands with coral fragments and cemented reefs or hard corals (Fig. 2.24). They are elevated successively with the ages from the sea levels, and denoted as older terraces towards inland and newer terraces along the shore fringe areas of the islands. There are three groups of terraces in Andaman Islands, e.g. coastal terraces, piedmont terraces and fluvial terraces, but reef terraces are found in the present islands. Seismo-tectonic activities and stages of earthquake cycles in the Late Quaternary period influence the development of reef terraces in the region. Reef terraces are not always well preserved in the islands due to weathering, erosion and surface modification processes under subaerial environment. Some reef terraces are monitored in the field since 2012 to identify their changes under tropical environment (Fig. 2.24).

The contour plans of ALOS PALSAR DEM helped in the present study to delineate the coastal reef terraces for the three islands. The elevated reef terraces of Neil Island are well preserved in 3, 5 and 10 mts heights above the sea level at present. Southwestern Sitapur has preserved a reef terrace at 3 mts elevation above the sea level, which comprises an area of cemented reef corals, with sharp-edged fingers. They are dominated by cup corals, finger corals and table corals at the shore fringes, and affected by wave erosion and weathering processes at an alarming rate (0.625mts/year).

Fig. 2.24 Eroded remnant of the younger reef terrace at Sitapur section of Neil Island

The surfaces of reef terrace are Karstified and produced many features like solution pipes, lappies, split karren, fissures, tunnels, arches on the sea face of the island coast. The coastal reef terraces of 5 m surface height above sea level comprising shell limestones with coral rags, sands and calcareous clays (Fig. 2.24). The porous surface is ideally providing natural recharge site of rain waters, and also provide areas for surface runoff in saturated condition. However, the shore fringe reef terrace of 10 mts elevated surface is consisting of fossiliferous corals, coral rubbles, compact bioclasts and bio-encrusted surfaces with karstified features in the island shore. The older terrace is partially covered by forest lands of matured logs, and exposed on the sea face by recession of shore cliffs in the region.

In Havelock Island, the coastal terraces of 5–10 mts elevated surfaces above sea level are widely preserved in northern and western shores. The Neil point and Radhanagar West preserved the residues of 3 mts elevated surfaces above sea level, but they are highly wave abraded at present, and bio-encrusted with rock boring animals and calcareous algae. Surface of the coastal terrace in the embayed shores of Havelock Island (northwestern part) exhibited karstified features with pinnacles and coastal karrens under the mats of grasses and calcareous algae.

The coastal fringe areas of Hut Bay, Harmindar Bay, and Buttler Bay of Little Andaman have well preserved the terraces of 3 mts, 5 mts and 10 m elevated surfaces on the eastern side fringed with the Andaman Sea. A 1.5 mts reef terrace is formed in 2004 by seismo-tectonic activities nearby the Jetty point of Hut Bay in Little Andaman. The raised beach terraces (1.5, 3, 5 and 10 mts) are also visible on the western shores adjacent of Jackson Creek and Bumila Creek, which are fringed with the Bay of Bengal. Raised corals with reefal rocks of 1.5 mts elevated surface are thinly covered by coralline sands and coral debris at present in the region. Woodlands of 5mts elevated surfaces are affected by the event of large-scale marine inundations (2004). A part of 10 m elevated reef terrace of Little Andaman nearby the Harmindar Bay was monitored since 2016–2019 in the field. The terrace with abandoned cliff comprised with compact reefal rocks, shell limestones, fossil debris of corals and taffoni marks under the sheds of coconut trees in the coastal region. Discontinuity in the surface morphometry also representing clear step-like surfaces in the island is generally described as resulting from uplift accompanying the events of large earthquakes (Lajoie, 1986; Berryman et al., 1989; Vita-Finzi and Hidayat, 1991; Ota and Yamaguchi, 2004). In Little Andaman Island, many drainage channels have produced waterfalls in their courses after crossing the break of slope of terrace fringed steps (at 20, 30 and 70 m contour heights). The surface uplift of 1.0–1.5 m above the tide level has produced a reef terrace after the event of 2004 earthquake in the shore fringe of Harmindar Bay. The coral reefs are also raised on the western shores (>1.0 m above sea level) of Little Andaman after the great earthquake in the region.

The age estimates of coastal terraces (radiocarbon ages of fossil coral terraces) of the Hut Bay areas by Rajendran et al. (2008) exhibit 620 ± 160 YBP to 7300 ± 300 YBP older successively from 1.7 to 4.5 m elevated surfaces on the coastal fringes. All the coastal terraces are located within the range of 1 km distance inland from the shorelines. The 4.5 m elevated terrace above the sea level was dated as 7300 YBP, and therefore estimated the rate of upliftment of the terrace as 1.62 mm/year. Such co-seismic elevation change along the coastal fringe has generated reef terraces in other islands of South Andaman (e.g. Havelock and Neil Island).

Shore platform mangroves are elevated in northern embayed shores of Havelock Island and in Kalapathhar areas of Little Andaman during the event of 2004 earthquake as evidenced in the field studies (2012–2016) of the present work. The terrace with coral rags and mangrove swamp clays are emerged in this way of upliftment along the shore fringes after the great earthquake (2004) in Andaman. Earlier studies (Rajendran et al., 2008, Das and Pramanik 2011, Bandopadhyay and Ghosh, 2015, Srinivasan and Chatterjee, 1981) related with the co-seismic uplift of Andaman shores after 2004 great earthquake reveal that the parts of North Andaman (Landfall Island, North Reef Island) and Middle Andaman (Interview Island) registered maximum upliftment (>1.5 m above m.s.l) than the South Andaman (0.70–1.0 m).

The raised beach terraces of Neil Island at Govind Nagar and Bharatpur at 3–4 m heights represent coral sands with coral rags and mangrove swamp clays of recent to sub-recent origin. Mangroves and littoral forests are adjusted in these terraces with the impacts of tidal pumping from the seaward sides and seepage flow of fresh waters from the landward sides along the shores (Plate 36).

2.2 Types of Wooded Island on a Reef Platform

2.2.1 Mangrove Ecosystems Over the Carbonate Platform

Shore platforms and raised beach platforms of the shore fringe areas, Ria-type estuary fringes, backwater fringes, tidal creek bank flats and coastal embayments with favourable environmental gradient supported unique ecosystem of mangroves under carbonate settings in all three islands of South Andaman district (Fig. 2.25).

2.2.2 Mangrove Ecosystems of the Intertidal Shore Platforms

There are three major physiographic settings for the mangroves habitat in the tropical coast. The islands of Andaman Sea and Bay of Bengal represent carbonate setting for mangroves in the tropical coast. Carbonate sedimentary environment of intertidal shore platforms raised beach platforms, Ria-type shores, backwater areas, other estuary channels provide environmental condition for mangrove growth in Little Andaman, Havelock Island and Neil Island (Fig. 2.26a). The source and supply of sediments usually determine the substrate condition of each geomorphic setting favourable for mangrove growth. Only the 'Ria' type shores and back waters with lagoonal settings have absorbed or accumulated organic nutrient-rich muds and carbonate clays transported from outside of the sheltered environment in the region. They are usually termed as allochthonous sediments and supported prolific growth

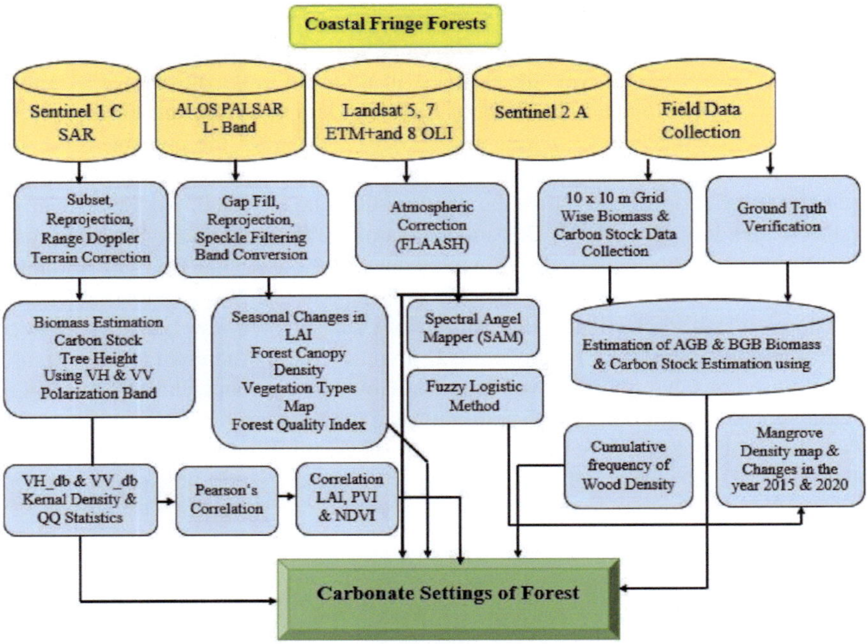

Fig. 2.25 Methodological sequences of the study

of mangroves. Fresh water inflow into the system by drainage channels from the island fringe hill slopes and tide water encroachment from the outside (marine environment) provided hydrologic conditions for mangrove growth under tropical hot and humid environment.

Other type of estuary sections linked with active drainage channels transport sediments loads by weathering and denudation of the island interiors, and also by tidal ingression into the channel mouths with sediment loads during dry season (November to March) in island territory. Number of drainage channels across the littoral tract in Havelock Island and Little Andaman supported mangroves along their lower sections. Radhanagar drainage channel of Havelock Island and Dugong River of Little Andaman belong to this category. Extensive growth of mangroves dominated by *Nypa fruticans, Rhizophora mucronata* (Fig. 2.26b), *Rhizophora apiculata, Bruguiera parviflora and Bruguiera gymnorrhiza* occurs along the intertidal flats and creek banks of Dugong River in Little Andaman. The river has drained the carbonate flat of eastern littoral tract over the limestone rocks, and produced a blanket of thick nutrient-rich organic muds with frequent exposures of sheety limestone in and around Netaji Nagar. Such a dense mangrove forest in the littoral tract accumulated tidal muds and decomposed organic matters or mangrove-derived peats in a large volume as autochthonous sediments (Fig. 2.26c). Tidal processes in the coastal fringes provided condition for advection of muds into the backshore wetlands

Fig. 2.26 Mangroves over the intertidal shore platforms in Havelock Island, **a** Mangroves of the embayed shore of Havelock island were affected by tsunami hazard during 2004 great earthquake, **b** Rhizophora Mucronata on the sea shore, **c** River fringe backwater mangroves (Dugong Creek) and **d** Shore fringe and river fringe mangroves (Neil and Little Andaman)

from the offshore during high tides in a considerable volume. The shore fringe sand dunes have separated the coastal wetlands from offshores other than the mouths of estuaries along the eastern coastal tract of Little Andaman.

Mangroves fringed 'Ria' type estuaries are abundant in northern littoral tracts of Little Andaman and Havelock Island in South Andaman. The drowned river valleys with wide valley flats, meandered tidal creeks and valley fringe-terraced banks provided conditions for storage of accumulated sediments transported from offshores in the region. The events of large-scale marine inundations of the geological and recent past (e.g. 2004 tsunami, cyclone surge of 2016, and frequent swell waves from south east) pushed sediments into the drowned valleys to develop wide shallow flats and tidal flats for the luxuriant growth of mangroves. Marine sediments with coral rags, bioclasts and mangrove litters are largely accumulated in such estuaries of the region. Dense mangrove forests occupy majority of the forest tract of the northern littoral plain in Little Andaman. The northern part of Havelock Island also occupies mangrove fringed 'ria' shores on the eastern margin of promontory. Both *Rhizophora sp.* and*Bruguiera sp.* are found on the banks of the drowned river valleys, but mixed vegetations of mangroves and saltmarsh tracts extend on the landward sides of such 'Ria' type shores in the region (Fig. 2.26d).

An identical site of backwater (Fig. 2.26d) is developed in southwestern part of Havelock Island over a large doline depression with volume of accumulated carbonate sediments. The backwater area is separated from Andaman Sea by barrier bars (covered by dense forests), and connected with the open marine environment with a tidal inlet. The wide basinal-shaped depression with sediment filled surface is fed by three different drainage channels from the landward sides, and the entire tidal basin is fringed with mangroves and saltmarsh vegetations at present. Large volumes of carbonate sands, silts and coral rags were deposited at the mouth during tsunami event (2004) in the western shores.

Raised beach platforms are another physiographic setting which also supports the environmental condition in the lithified carbonate sediments for mangroves. The porous sediments of the raised beach platforms at the backshores also supported condition for existence or survival of mangroves in many areas of Havelock and Neil Island. Landward fringes of such raised beach platforms are lined with littoral forests or mangrove associates in the tropical coast. Holocene reef flat was raised along the coastal fringes (around 2000 YBP-4000 YBP) in the archipelago group of islands (Fig. 2.27).

The seismo-tectonic impacts of 2004 mega earthquake (RH 9.1) have generated such types of raised beach features also along the coastal fringes of these islands. Supratidal mangroves and upper intertidal mangroves are sufficiently raised in the sites of Kalapaththar (Little Andaman), Elephant beach (Havelock Island), Govindnagar (Neil Island) and Dugong creek mouth (Little Andaman) in South Andaman. Areas under *Bruguiera gymnorhiza, Kandelia candel, Acrostichum aureum and Lumnitzera racemosa a*re mostly affected by tectonic upliftment.

Carbonate shore platforms at the coral fringed islands of South Andaman created environmental gradients for growth and regeneration of mangroves. Island fringe shore platforms and Holocene reef terraces of Havelock Island, Neil Island and

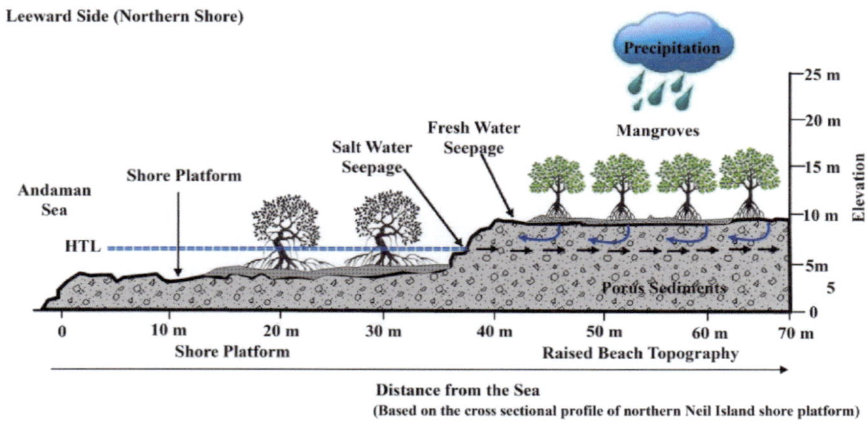

Fig. 2.27 Physiographic setting for mangroves in raised beach topography with composition of porous sediments of carbonate materials (Neil Island, Govindnagar section)

Little Andaman supported mangrove forest stands on the sea face. Most of the shore platforms are backed by sea cliffs and fronted by reef terrace and reef slope on the windward sides in the region. Sub-horizontal platforms and sloping platforms of these islands have developed on resistant rocks and relatively weak rocks in different sections of the shoreline configurations. There is a strong relationship between shore platform gradient and tidal range but differences in slope indicated the Holocene sea level history and the significant role of weathering and wave actions of the past.

Sediments are mostly supplied from the reef slopes (reef rocks, boulders, gravels of carbonate origin) and also from the landward sides (rock falls, debris falls and soluble materials) of the island. Waves and tides transport graveliferous and shingle size–sand size sediments, and finer sediments on the platforms on the basis of slope and tidal range. Accumulated sediments are once colonized by mangroves the regeneration process continues to spread over the region in extensive forest tract and provide conditions for further accumulation of carbonate silts, sands and muds by repeated tidal inundations.

Shore platforms of the embayed shores, sub-horizontal surface and accumulated sediments are colonized by dense mangrove forest in Havelock Island, Neil Island and Little Andaman on the sea face. Mangroves of the shore platforms are dominated by *Rhizophora mucronata, Bruguiera gymnorrhiza, Kandelia candel, Rhizophora apiculata, Avicennia officinalis and Bruguiera parviflora* in this island of South Andaman (Fig. 2.26). The substrate is generally peaty in which organic matters accumulated by mangroves and only limited inorganic sediments transported from outside by marine inundations. Repeated visit to the sites of shore platform mangroves also helped to identify the favourable conditions and environmental stress over the residual forests. The fresh water seepage lines and seepage pits make the surface as mixing zone of salt water-fresh water complex on the intertidal shore platforms for favourable condition of mangrove ecosystem development in many parts of Neil Island and Havelock Island. However, the events of large-scale marine inundations by cyclones and tsunami incidence (2004) have generated environmental stress for damaging mangroves in the shore platforms.

2.2.3 Distribution of Zoned Mangroves

Mangroves of the wet tropics and near equatorial environments are well-grown vegetation and they have extended into the zones of rainforest communities along the shores of Andaman Islands. Mangroves are also 'zone' species over the carbonate sediments and indicating the adaptive responsive with presence of strong environmental gradients across the tidal zones of the region. Along the intertidal zones of Dugong creek and other Ria-type estuaries of Little Andaman, mangroves are widely distributed over the landward fringes of tidal flats and they have been mixed with littoral forests and rainforests on the elevated surface. Dense canopy covers and intricate roots of the mangrove are found along the sheltered embayed shores of Havelock

Island. Mangroves of the carbonate setting also have interacted with tropical rainforests by adapting the long and tall growth (over 40 m), luxuriant growth, buttressed trunks, wide canopy covers, growth of lianas and other epiphytes in the region of strong hydrological cycle. Andaman mangroves are also showing then great diversity in species composition (over 32 species), particularly dominated with *Rhizophora sp., Avicennia sp., Bruguiera sp., Lumnitzera sp. and Nypa sp.* along the tidal zones. Physiological tolerance depends on the local inundation regimes, soil anoxia, water salinity, nutrient loads, freshwater-saltwater mixing zones and nature of carbonate sediments available in the oceanic islands (Fig. 2.28).

Mangroves have expanded landwards into the sedge lands along the backwater fringes of Havelock Island and Neil Island in which long and tall grasses are densely distributed over the carbonate muds and become affected by large events of marine inundations (Fig. 2.29). Mangroves swept away from the backwater fringes by tsunami waters and external inorganic sediments input after the event of subsidence caused by seismo-tectonic activities during 2004 earthquakes along the western shores of Havelock Island. The study reveals that mangroves of such carbonate settings adjust with morphological resilience, tectonics, hydrodynamics, sediment transport, sea level fluctuations and episodic energy inputs. The succession

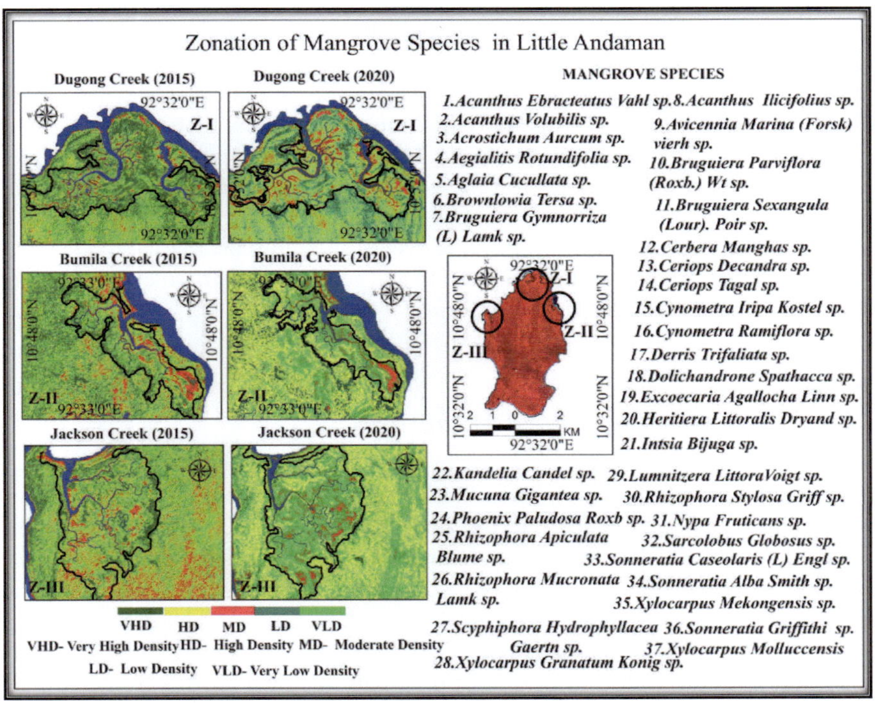

Fig. 2.28 Mangrove species distribution in different zones of Little Andaman

Fig. 2.29 Kandelia Candel species on the shore platform (Havelock Island)

patterns of mangroves are also reflected by physiographic settings and diversity of morphological resilience's in the island shores of South Andaman.

The shore platform mangroves are fringe with littoral forest on the landward side and mixed with semi-evergreen to evergreen forests further inland. Tidal woodland forests of estuaries and 'Ria' shores however, backed by sedge plants and semi-evergreen forests in the landward parts (Figs. 2.29 and 2.30). The littoral forest is absent in the region of raised platforms and moist tidal flats along the landward fringes.

Mangroves are distributed in patchy forms in the shore platforms and raised beach platforms. Areas of seepage waters and accumulation of finer sediments into the carbonate shore platforms exhibited the locational pattern of mangroves in the shore fringe rocky tidal flats in the oceanic islands. Extensive field studies and remote sensing studies helped to estimate the mangrove patches in the shore platforms and over the raised beach platforms (Table 2.8 and Fig. 2.31).

The Neil Island and Havelock Island constitute major patches of mangroves on the shore platforms. Havelock has 56 patches, and Neil Island has 48 patches of mangroves along the shore platforms. However, Little Andaman Island consists only 29 patches of mangroves on its shore platforms (Table 2.9 and Fig. 2.30). Mangroves are also distributed in the raised beach platforms of the islands. Mangroves-dominated shore platforms of swampy clay with coral rags were uplifted into supratidal flat after the event of seismo-tectonic activities (2004) along the shorelines of Little Andaman. Such mangroves of the raised beach platforms are existing in many places, in which the habitat is modified and adjusted with the tidal pumping and seepage water flows under carbonate settings. Mangroves of the raised beach platforms are located on the western and eastern shores of Little Andaman, north-eastern shores of Neil Island, and Elephant beach areas of Havelock Island (Paul et al., 2017, Paul et al., 2018a, Paul et al., 2018b). Other areas of mangrove habitats are situated along the estuary fringes, creek margin flats and basin fringe areas of the carbonate islands.

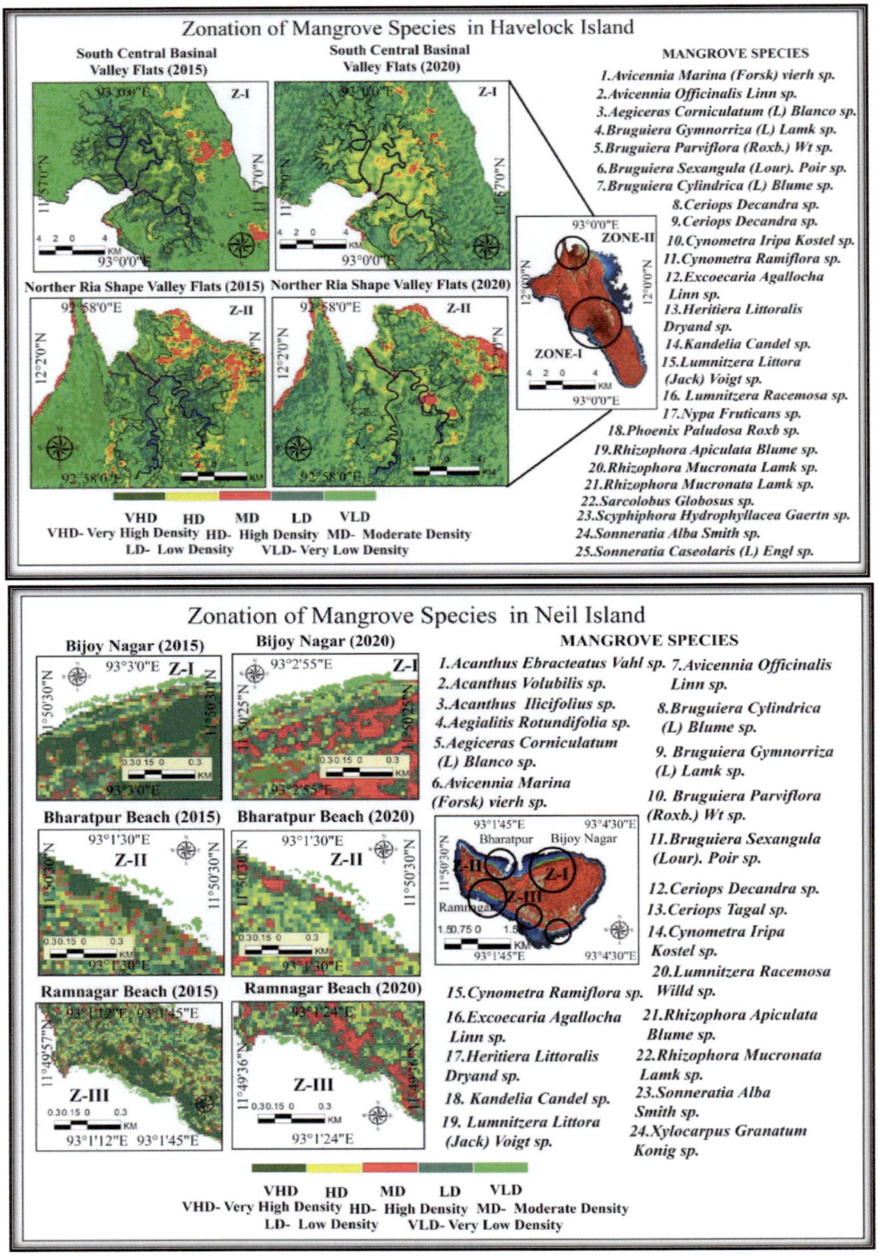

Fig. 2.30 Mangrove species distribution in Havelock and Neil Island

Table 2.8 Mangrove habitats on the carbonate settings under shore platforms and shore fringe areas of the three islands

Name of the Islands	Mangrove Patches (shore platforms)	Mangrove patches in other part
Neil Island	Jetty point patches-10 Lakshmanpur West-03 Neil Kendra patches-05 Ramnagar patches-10 Atreya mangrove point-07 Govind Nagar patches-12	Raised beach platform patches
Havelock Island	Radhanagar West- 5 Radhanagar South-West-11 Elephant beach-08 Northern Embayment-09 Govidnagar-06 Kalapathar South East-17	Northern Ria Estuary Fringe Mangrove-06 South Central Basin Fringe Areas-06
Little Andaman	Bala Hanuma Beach-04 Little Andaman West Bay- 08 Little Andaman West-04 East Bay Patches-04 South Bay Patches-09	Dugong Creek Fringes-20 Northern Ria Estuary Fringe-20 Mangrove on the Raised Beach Terraces-18

Fig. 2.31 **a** Residual mangroves on the wave-abraded shore platform (Neil Island), **b** Upper high tide shore platform mangroves with anchored root patterns

2.2.4 Favourable Environmental Gradient for Mangrove Habitat

The substrate conditions for mangrove habitats in the carbonate settings exhibit organic sediments of coral reef debris, sands and carbonate clays with other bioclasts along the intertidal shore platforms. Partial lithification and dynamic accumulation of organic sediments may produce hindrances to arrest and to anchor the mangrove propagules in the open marine environment. Only few mangrove seeds can survive in such substrate conditions (e.g.*Rhizophora, Bruguiera species*) to develop colony of mangrove habitats. Survival of mangroves under different carbonate settings depend

Table 2.9 Species diversity of mangroves in Havelock, Neil and Little Andaman Islands (species distribution shown in green colour)

Mangrove species (Havelock Island, Neil Island and Little Andaman)				
Sl. no.	Species name	Havelock Island	Neil Island	Little Andaman
1	*Acanthus Ebracteatus Vahl sp.*	✓	✓	✓
2	*Acanthus Ilicifolius sp.*	✓	✓	✓
3	*Acanthus Volubilis sp.*		✓	✓
4	*Acrostichum Aurum sp.*			✓
5	*Aegialitis Rotundifolia sp.*		✓	✓
6	*Aegiceras Corniculatum (L) Blanco sp.*	✓	✓	
7	*Aglaia Cucullata sp.*	✓		✓
8	*Avicennia Marina (Forsk) vierh sp.*	✓	✓	✓
9	*Avicennia Officinalis Linn sp.*	✓	✓	
10	*Brownlowia Tersa sp.*			✓
11	*Bruguiera Cylindrica (L) Blume sp.*	✓	✓	
12	*Bruguiera Gymnorriza (L) Lamk sp.*	✓	✓	✓
13	*Bruguiera Parviflora (Roxb.) Wt sp.*	✓	✓	✓
14	*Bruguiera Sexangula (Lour). Poir sp.*	✓	✓	
15	*Cerbera Manghas sp.*			✓
16	*Ceriops Decandra sp.*	✓	✓	✓
17	*Ceriops Tagal sp.*	✓	✓	✓
18	*Cynometra Iripa Kostel sp.*	✓	✓	✓
19	*Cynometra Ramiflora sp.*	✓	✓	✓
20	*Derris Trifaliata sp.*			✓
21	*Dolichandrone Spathacca sp.*			✓
22	*Excoecaria Agallocha Linn sp.*	✓	✓	✓
23	*Heritiera Littoralis Dryand sp.*	✓	✓	✓
24	*Intsia Bijuga sp.*			✓
25	*Kandelia Candel sp.*	✓	✓	✓
26	*Lumnitzera Littora (Jack) Voigt sp.*	✓	✓	✓
27	*Lumnitzera Racemosa Willd sp.*	✓	✓	
28	*Mucuna Gigantea sp.*			✓
29	*Nypa Fruticans sp. Thunb, Wurnb*	✓		✓
30	*Phoenix Paludosa Roxb sp.*	✓		✓
31	*Rhizophora Apiculata Blume sp.*	✓	✓	✓
32	*Rhizophora Mucronata Lamk sp.*	✓	✓	✓
33	*Rhizophora Stylosa Griff sp.*	✓	✓	✓
34	*Sarcolobus Globosus sp.*	✓		✓
35	*Scyphiphora Hydrophyllacea Gaertn sp.*	✓	✓	✓
36	*Sonneratia Alba Smith sp.*	✓	✓	✓
37	*Sonneratia Caseolaris (L) Engl sp.*	✓	✓	✓
38	*Sonneratia Griffithi sp.*			✓
39	*Xylocarpus Granatum Konig sp.*	✓	✓	✓
40	*Xylocarpus Mekongensis sp.*			✓
41	*Xylocarpus Molluccensis sp.*			✓
	Total Species	**28**	**27**	**36**

on environmental gradient. Various physical, biological, hydrological, chemical and sedimentological conditions available in different settings support mangrove habitats in such carbonate islands. As a consideration of substrate conditions with intertidal environment, thick sediment deposits, sheltered location, available ground for dispersal of seeds, fresh water salt water mixing environment and favourable salinity front with nutrient loads, the 'Ria' type estuaries or drowned valleys and other estuarine drainage channels of Little Andaman and Havelock Island support luxuriant growth of mangroves. The fresh water sources, sediment supplies and tidal inflows into the low gradient sections of estuaries and drainage channel mouths made the coastal system favourable for exchange of sediment, water and nutrient loads. Finer sediments are accumulated on the tidal flats of Ria-type estuaries and along the lower section of longer drainage channels. Salinity fluctuates during rainy months (April to October) and in dry season (November to March) of a year with discharge variations in the streams. Drainage channels of the eastern littoral tract of Little Andaman flowing towards the Hut Bay and Buttler Bay are fringed with *Nypa fruticans and Rhizophora apiculata* which indicate the availability of fresh water supply into the downstream section (Table 2.9 and Fig. 2.31).

All the river mouths of the region are raised and filled up with inorganic sediment transport after the events of earthquake and tsunami (2004) activities. The backshore water bodies of such ponded stream beds are now transformed into backwater areas with hyper-saline condition, and causing significant changes in species diversity along the coastal section of streams. General environmental gradient of the shore platform mangroves is changed due to dumping of large boulders, inorganic sediments, calcareous reef limestone debris and other bioclasts after the events of storms and tsunami inundations. Mangrove propagules and other types of viviparous seeds are not arrested or anchored in the current substrate condition under increased rate of wave abrasion (Fig. 2.31).

Thus, mangrove regeneration process is affected in such unfavourable condition of the shore platforms in South Andaman Islands. Salinity of sea water is very high (25–27 ppt) as recorded in different sites of mangrove-dominated shore platforms, and the water pH was ranging from 7.5 to 9.2 in the same areas under residual mangrove stands. Areas of shore platforms nearby embayments and headlands only provided condition for finer sediment accumulation to support mangrove colonies. Peaty soils or organic matters are generated in the places of dense forest of sheltered tract along the shore platforms.

A large backwater setting is located in the western coastal tract of Havelock Island behind the barrier spits in which salinity of water achieved up to 30 ppt in the month of January, 2016 and water pH was ranging from 7.2 to 8.00 under protective environment. High salt-tolerant mangroves and salt marsh vegetations with *Sargassum* species are found along the banks of inlet channel. Accumulation of carbonate clays on the landward margins linked with three drainage channels of the basinal depression supported dense halophytic grassland with dwarfed mangroves under euryhaline condition.

Environmental gradient for mangroves on the raised beach surface created a new setting in which tidal inundations are largely reduced except the rare events of marine inundations in the region (Havelock Island and Neil Island) by tsunamis and cyclonic storms surges. Tidal pumping at the high tides created condition of salt water encroachment through percolation process into the porous materials of raised beach areas to support the growth of mangroves in such elevated surfaces. Low-diversity species of vegetation complex exist in the interface of land and sea in comparison with the estuaries, embayments and lagoons of the tropics. Mangroves of the region are dominated by *Bruguiera gymnorhiza, Acrostichum aureum and Excoecaria agallocha* (Fig. 2.31).

2.2.5 Regeneration of Mangroves Over the Carbonate Platform

Mangroves have attained their maximum development and significant diversity along moist warm tidal flats of estuaries and embayments of the tropical island coasts. Many mangroves like *Rhizophora, Bruguiera and Ceriops* are viviparous in character or the seeds germinate still attached to the canopy of trees (Rabinowitz, 1978). Dispersal of the torpedo-shaped propagules on the substrate by tidal currents also appears to be correlated with the pattern of mangrove zonation under wide tidal flats dominated by finer sediments. Protected and sheltered areas of tidal flats only can support usual environment for accumulation of sediments and regeneration of mangroves by anchoring the germinated propagules over them. There are some frontal mangroves like *Avicennia officinalis, Avicennia marina* which can produce seedlings of another category that can withstand for a long time on the tidal flats due to their ability to respire anaerobically along the embayed shores of island platforms (Table 2.10, Fig. 2.32).

Various types of truncation process, beach rock formation by cementing the calcareous materials, dumping of large boulders by marine inundations and by transporting the inorganic sediments over the island fringe shore platforms produced environment of unfavourable condition against the mangrove regeneration process at present in South Andaman Island. Modern sea level rise process with frequent cyclones transport the inorganic sediments into the shore platforms to disturb the regeneration process of mangroves. Only few propagules can withstand the harsh environment on the shore platforms due to emerged boulder ridge surface after tsunami incidence (2004), and significant wave breakers during rising tides along the shorelines of such oceanic islands in the region (Table 2.10). Thus, Andaman shore platform mangroves are significantly reduced at present in Havelock Island (Bijoy Nagar, Shyam Nagar, and Krishna Nagar), and Neil Island (Neil Jetty Point, Govind Nagar, Ramnagar, Neil Kendra, and Lakshmanpur-I) after the events of large-scale marine inundation (2004, 2016).

The pneumatophores (breathing roots) of *Avicennia sp.* are trapped with inorganic sediments over the shore platforms in many places after repeated cyclone breaks in South Andaman Islands. Mangroves seeds have been decayed and decomposed on the surface of accumulated inorganic sediments in the region. Regeneration of new colony of mangroves was restricted due to above reasons after such episodic energy outbreaks in the island systems. Areas of mangroves (*Rhizophora sp., Bruguiera sp., Avicennia sp.*) of the intertidal shore platforms are raised and

Table 2.10 Changing mangrove cover areas in the islands

Year	Total area of mangrove (Hectares)		
	Havelock Island	Neil Island	Little Andaman
2015	879	291	2353
2020	912	250	1807

Fig. 2.32 Mangrove roots of different environmental gradient on the shore platforms

moved into the supratidal zone at western and eastern shores of Little Andaman due to earthquakes or seismo-tectonic activities. Mangrove regenerations by dispersal of seeds are abandoned at present due to events of the rare inundations in the raised platforms (Fig. 2.33).

Fig. 2.33 The changing mangrove cover areas between 2015 and 2020

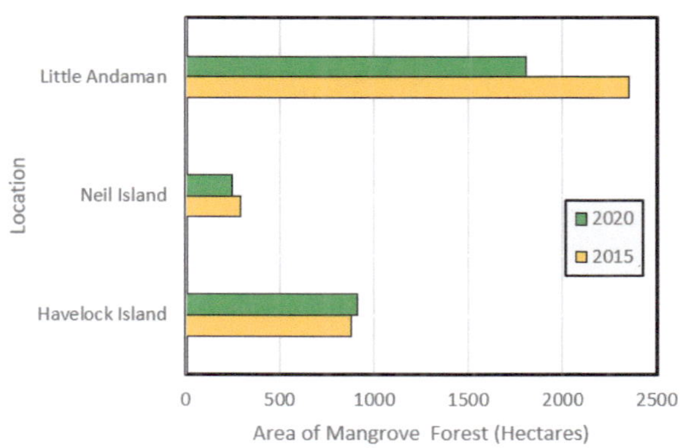

2.2.6 *Physiological Adaptations of the Mangroves on the Carbonate Platforms*

Mangroves in the soft sediment and carbonate sediment have adapted differently by extending cable roots and surface roots in the intertidal region of the island shores (Tables 2.10 and 2.11, Figs. 2.34 and 2.35). The underground roots are richly supplied with finer anchor and absorptive roots particularly in the soft sediment surface of 'Ria' type shores and inland parts of estuaries. Above ground roots of different categories include pneumatophores *(Avicennia sp.)*, prop roots *(Rhizophora sp.)* buttresses and knee roots *(Bruguiera sp.)* perforated breathing roots *(Phoenix sp.)* root-stock prostrate *(Nypa sp.),* stilt roots with lianas *(Rhizophora apiculata),* laterally spreading snake like roots *(Excoecaria sp.)*.

Table 2.11 Statistical calculation of wood density based on the field data of the three islands

Sl. no	Species name	Mean	Median	Mode	SD	Variance
1	*Acanthus Ebracteatus Vahl sp.*	0	0	0	0	0
2	*Acanthus Ilicifolius sp.*	0	0	0	0	0
3	*Acanthus Volubilis sp.*	0	0	0	0	0
4	*Acrostichum Aurcum sp.*	0	0	0	0	0
5	*Aegialitis Rotundifolia sp.*	0	0	0	0	0
6	*Aegiceras Corniculatum (L) Blanco sp.*	0.5967	0.64	0.64	0.0751	0.0056
7	*Aglaia Cucullata sp.*	0.6209	0.62	0.77	0.1197	0.0143
8	*Avicennia Marina (Forsk) vierh sp.*	0.7316	0.7	0.81	0.1547	0.0239
9	*Avicennia Officinalis Linn sp.*	0.6987	0	0.83	0.1217	0.0148
10	*Brownlowia Tersa sp.*	0	0	0	0	0
11	*Bruguiera Cylindrica (L) Blume sp.*	0.81	0	0.72	0.1273	0.0162
12	*Bruguiera Gymnorriza (L) Lamk sp.*	0.8683	0.94	0.94	0.1103	0.0122
13	*Bruguiera Parviflora (Roxb.) Wt sp.*	0.8427	0	0.92	0.0799	0.0064
14	*Bruguiera Sexangula (Lour). Poir sp.*	0.83	0	0.74	0.1273	0.0162
15	*Cerbera Manghas sp.*	0.44	0.47	0.47	0.052	0.0027
16	*Ceriops Decandra sp.*	0.725	0	0.79	0.2453	0.0602
17	*Ceriops Tagal sp.*	0.08859	0	0.883	0.0821	0.0067
18	*Cynometra Iripa Kostel sp.*	0.8411	0.84	0.8415	0.116	0.0135
19	*Cynometra Ramiflora sp.*	0.8415	0.88	0.93	0.0951	0.009
20	*Derris Trifaliata sp.*	0	0	0	0	0
21	*Dolichandrone Spathacca sp.*	0.405	0.4	0.56	0.0951	0.0091
22	*Excoecaria Agallocha Linn sp.*	0.4288	0.45	0.49	0.0557	0.0031
23	*Heritiera Littoralis Dryand sp.*	0.8847	0	0.98	0.1085	0.0118

(continued)

Table 2.11 (continued)

Sl. no	Species name	Mean	Median	Mode	SD	Variance
24	*Intsia Bijuga sp.*	0.84	0	0.84	0	0
25	*Kandelia Candel sp.*	0.5273	0	0.58	0.053	0.0028
26	*Lumnitzera Littora (Jack) Voigt sp.*	0.727	0	0.67	0.0705	0.005
27	*Lumnitzera Racemosa Willd sp.*	0.8325	0	0.88	0.0822	0.0068
28	*Mucuna Gigantea sp.*	0	0	0	0	0
29	*Nypa Fruticans sp. Thunb, Wurnb*	0	0	0	0	0
30	*Phoenix Paludosa Roxb sp.*	0	0	0	0	0
31	*Rhizophora Apiculata Blume sp.*	0.8814	0.85	1.05	0.2097	0.044
32	*Rhizophora Mucronata Lamk sp.*	0.8483	0	0.904	0.1013	0.0103
33	*Rhizophora Stylosa Griff sp.*	0.94	0	1.04	0.1414	0.02
34	*Sarcolobus Globosus sp.*	0	0	0	0	0
35	*Scyphiphora Hydrophyllacea Gaertn sp.*	0.685	0	0.69	0.0071	0
36	*Sonneratia Alba Smith sp.*	0.6443	0	0.78	0.1855	0.0344
37	*Sonneratia Caseolaris (L) Engl sp.*	0.5337	0	0.62	0.1624	0.0264
38	*Sonneratia Griffithi sp.*	0	0	0	0	0
39	*Xylocarpus Granatum Konig sp.*	0.6721	0.663	0.7	0.136	0.0185
40	*Xylocarpus Mekongensis sp.*	0.6535	0	0.74	0.089	0.0079
41	*Xylocarpus Molluccensis sp.*	0.6535	0	0.74	0.0891	0.0079

Fig. 2.34 Matured mangroves on the bank margin of northern embayed coast of Havelock Island

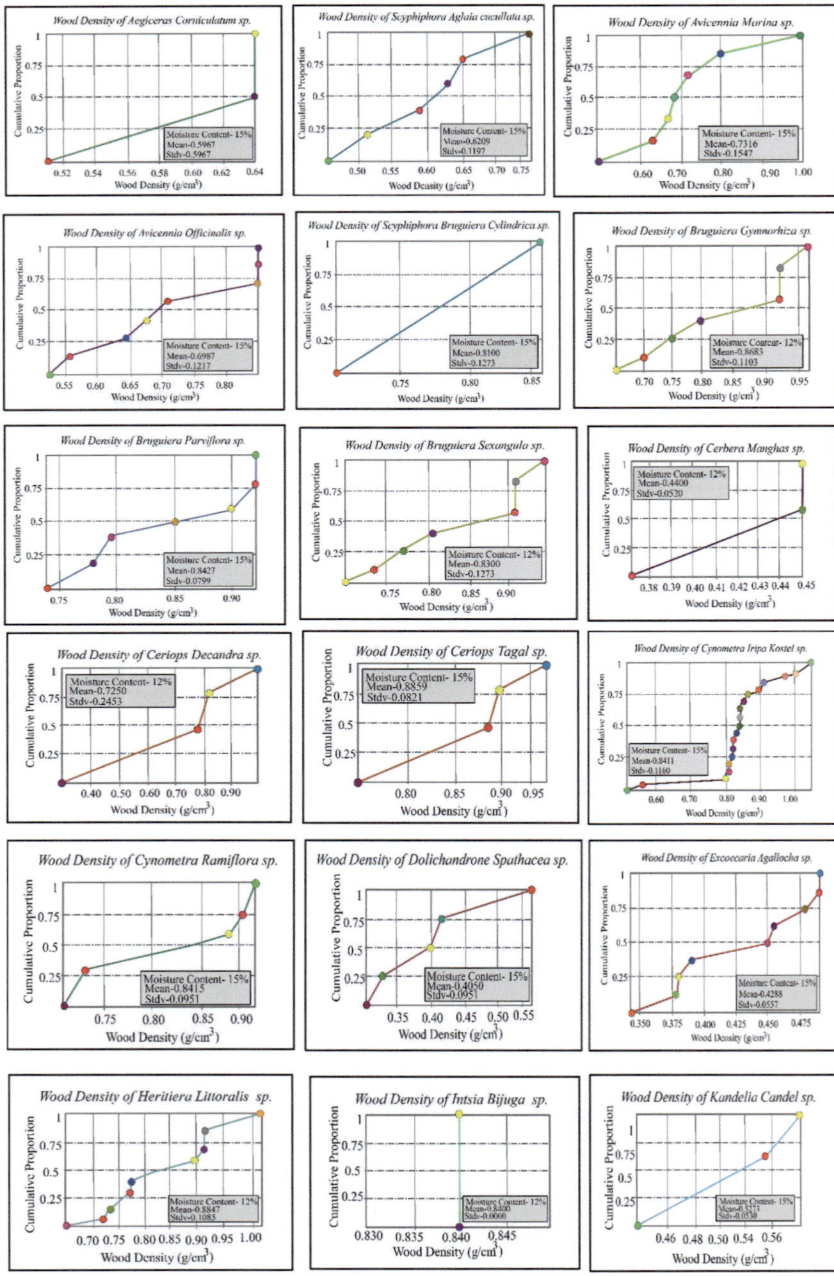

Fig. 2.35 Cumulative distribution of wood density of the different mangrove species (Island systems)

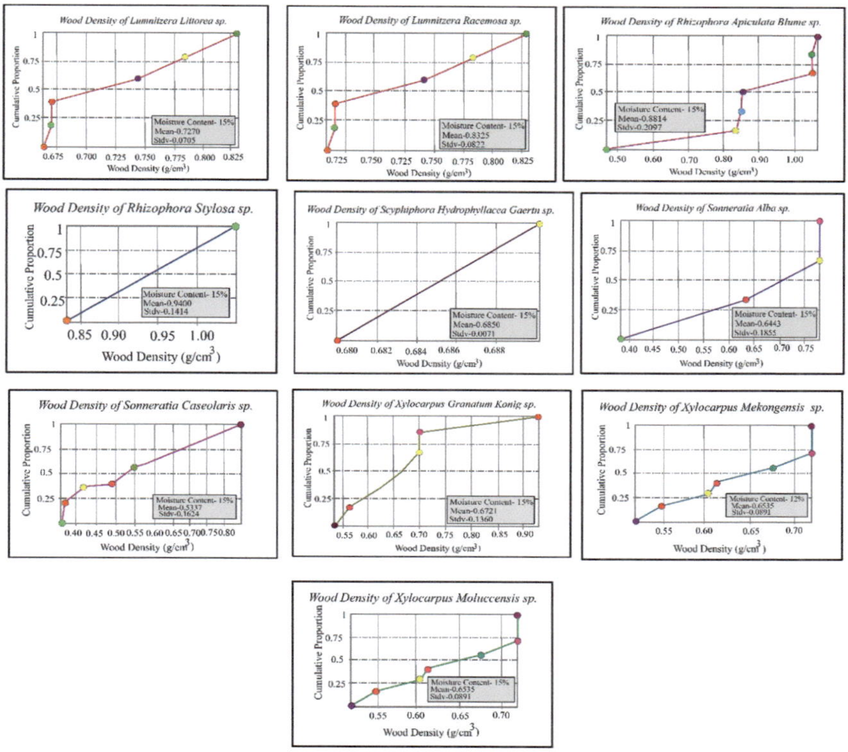

Fig. 2.35 (continued)

2.2.7 *Wood Density of Mangroves*

Wood density varies from species to species and their growth maturity with locational diversity in the shore platforms. The carbonate settings of mangroves in the sheltered areas may achieve optimum growth. Thus, wood density of mangroves of the Havelock Island and Neil Island represented adjustment of vegetations to the local substrate condition and available environmental gradient on the shore platforms. A few species like *Rhizophora apiculata, Bruguiera parviflora, Lumnitzera littorea, Avicennia marina and Sonneratia alba* are highly adjusted to the local environment and showing high wood density ranging from 0.78 to 1.05 g/cm^3 in the islands of carbonate setting. The graphical representation of cumulative percentage distribution of wood density is also showing variation from species to species in mangroves (Figs. 2.34 and 2.35, Table 2.11).

2.2.8 Other Forest Types of the Island System

The forest types of the carbonate platform of the islands are described in this section other than the areas belong to the shore platforms and estuary mouths. Various types of the forests are highlighted with their assessment of Leaf Area Index as well as their changing forest cover areas in this part of the chapter.

2.2.8.1 Other Forest Types

The Central Agricultural Research Institute (ICAR) for Andaman and Nicobar Islands classified the forest resources (1987) of the group of Islands. About 90% areas of Little Andaman are densely forested but Havelock Island and Neil Island still occupy 80% areas of their forest lands in South Andaman. There are 9 groups of forests in these islands: (i) Tidal Swamp Forests (Mangroves), (ii) Littoral Forests, (iii) Andaman Moist Deciduous Forests, (iv) Andaman Semi-evergreen Forests, (v) Bamboo Brakes, (vi) Cane Brakes, (vii) Southern Hill Top Tropical Evergreen Forests, (viii) Andaman Tropical Evergreen Forests and (ix) Giant Evergreen Forests (Fig. 2.36).

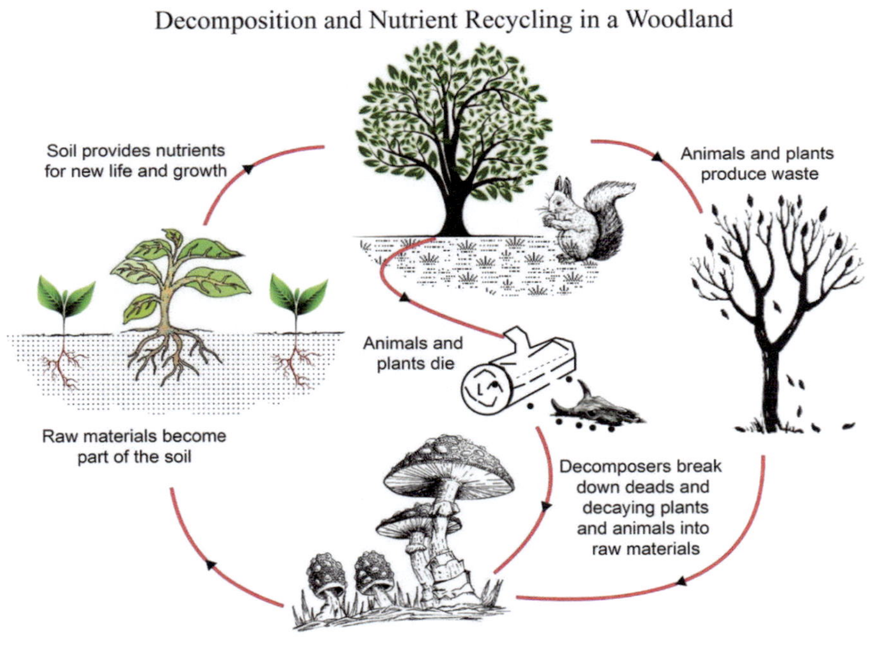

Fig. 2.36 Forest material decomposition and nutrient recycling process maintain the natural growth of the forest ecosystem

The tropical oceanic islands of South Andaman with average annual rainfall of 3087 mm, 27–30 °C annual temperature range, over 145 annual rainy days and 80% relative humidity with carbonate rocks, limestones and moisture soils support luxuriant growth of tropical rainforests. Among the above group of forests, Andaman Moist Deciduous, Andaman Semi-evergreen, Andaman Tropical Evergreen and Giant Evergreen trees are of major commercial value. The tropical rainforest trees (semi-evergreen, tropical evergreen and giant evergreen) are usually distributed over the hill tops, upper hill slopes and also on the raised reef terraces in these carbonate islands. Such rainforest trees have characteristic significant diametric growth (1.34–2.59 cm), long or tall stands (40–60 m), wide canopy covers (250–320 m^2), buttressed trunks (over 8 m tall and having 2.5 m gap between two buttresses), number of epiphytes, lianas, thick barks and dense shades over the island surfaces.

Rainforest trees with their wide canopy space, extended stems or branches, trunks, roots, diverse epiphytes and thick turf of organic matters and nutrient-rich soils support habitats for insects, multiple groups of animals and avifauna in the island systems of oceanic environment.

Field visits into the several sites of forest floors of Little Andaman and Havelock Island also helped to explore the nature of soil covers, root penetration into the jointed limestones, moisture variations, micro climate conditions and light energy distributions under the tree shades. The thickness of litters or the humus on the soil floor ranges from 15 to 30 cm, and deep penetration of the root system into the jointed rocks produced able conditions for transmission of rainwaters into the disintegrated limestone rock bodies for development of liquid bi-carbonate through solution process and nutrient-rich soils with 60–80% soil moisture content supported dense under growths of thickets in the region (Figs. 2.36 and 2.37, Table 2.12).

In many areas of dense forests, sunlight can penetrate only during the overhead situation of sun at 90° angles in the day time. The rainforests of these oceanic islands also can regulate their own hydrological cycle by releasing moistures through evapotranspiration rate into the year for saturating clouds to produce the local rains. Such self-sustaining ecosystem of the evergreen trees made it possible to hold the soil nutrients on the forest floor of sloping and flat surfaces into these islands by increased infiltration rate and also by reduced surface runoff during rains.

Forests have been removed from the hill slopes and intermontane valley flats by the people in Havelock Island and Neil Island for converting them into plantation gardens and also for producing paddy crops, to supply cash crops and to meet the local demands in South Andaman district. Fertility of the soil is reduced for such agricultural uses of the previous forest lands in these islands dominated by acid sulphate soils. Soil erosion and nutrient losses are enhanced with seasonal exposures of soils on the sloping lands and cultivation of such sensitive soils along the intermontane valley flats. The lowland valley soils of these islands have a pH in the range of 5.0–5.6, and upland areas have 4.0–5.0, but acid sulphate soils of a few areas have only 3.0–4.0 pH (Figs. 2.38, 2.39 and 2.40; Table 2.12). As the soils of upland areas are cultivated and open, the available organic carbon ranges from 0.8 to 1.3%, but the acid sulphate soils have organic carbon between 3.0 and 4.5% in the region.

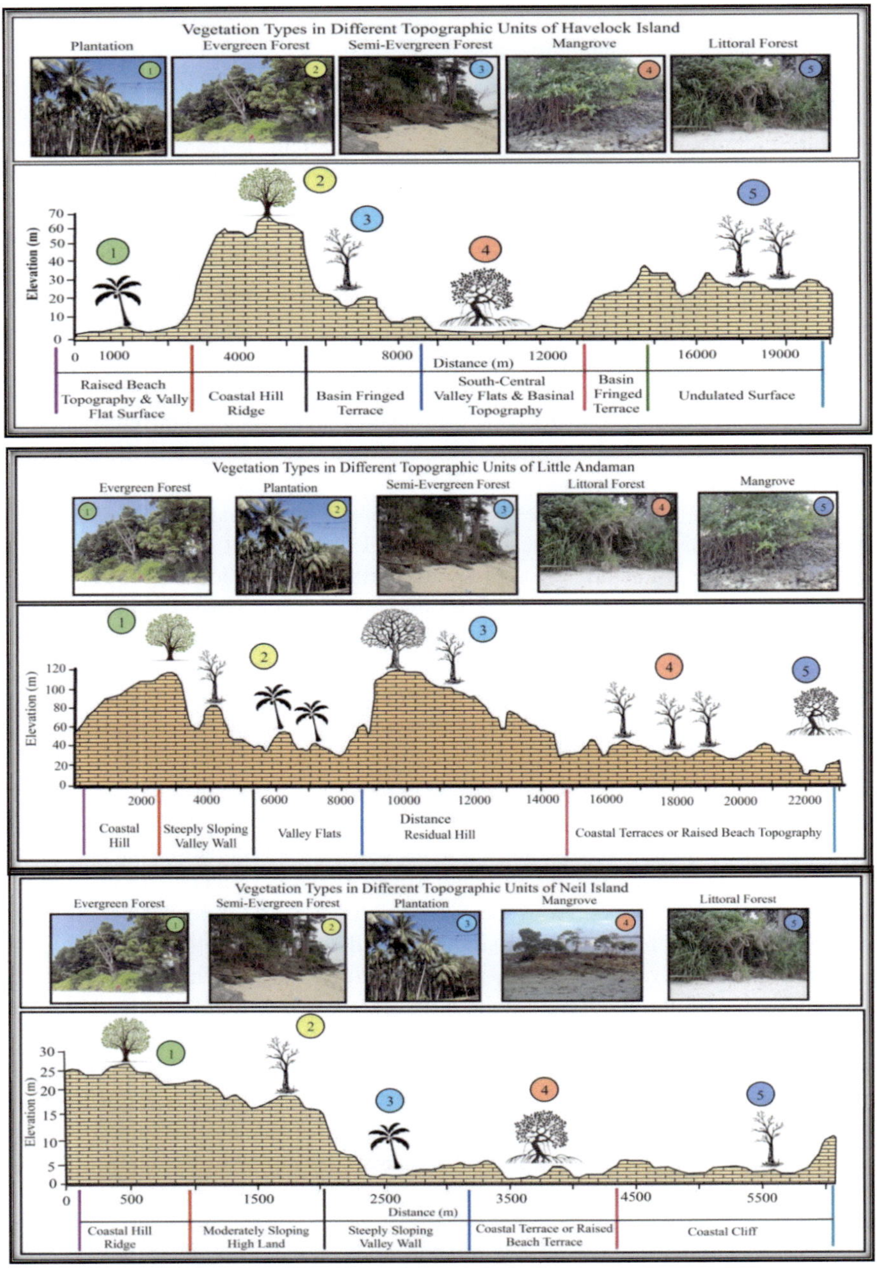

Fig. 2.37 Topographic vegetation model relationship

Table 2.12 Types of forests, major trees, ecology and physiography of the islands

Type of forest	Major trees	Ecology	Physiography
Andaman Semi-evergreen forest	Gurjan	Loftiest Tree (55–60), chief top storey tree, yellowish flowers	Pedimented surface of the hill slopes
	Letkok	Tall buttressed tree (45 m), fruits of the size like coconut	
	White Chuglam	Thin curved buttresses, butterfly shaped fruits (47 m)	
	Black Siris	Excellent Furniture Wood, (> 40 m tall)	
	Padauk	White colour trunk, large burrs on trunk (> 40 m tall)	
	Pyinma	Leaves turn reddish when falling, durable canoe wood, 35–40 m tall	
Andaman moist deciduous forest	Safed Chuglam	Wood smells of vinegar, resin used	Small stripes along the shore fringe terraces
	Red Silk cotton tree or Dida	400 stamens and larger petals	
	Jungly dungy	Grows to 45 m tall with enormous buttresses, and used for dug-out canoes	
	Marble wood	Wood figured with jet black stripes	
	Drinking cane	Climbing over tall trees	

The available data from Central Agricultural Research Institute (CARI) of Port Blair (1986) shows that the total geographical areas of forests for Neil Island were 676 ha, Havelock Island were 9903 ha and Little Andaman were 70,385 ha respectively in South Andaman district. However, the remote sensing studies of the present work (2018) showed change in geographical area of forest covers in these climate-sensitive carbonate islands of South Andaman (Table 2.12).

Andaman Semi-evergreen and Andaman Moist Deciduous forests are located on the elevated surfaces of shore fringed terraces and also along the pedimented surfaces around linear ridges in these islands. Such types of forests were surveyed along the reef terraces of South Bay, Harminder Bay, Hut Bay and Buttler Bay in Little Andaman; Lakshmanpur, Neil Kendra, Ramnagar and Govindnagar areas of

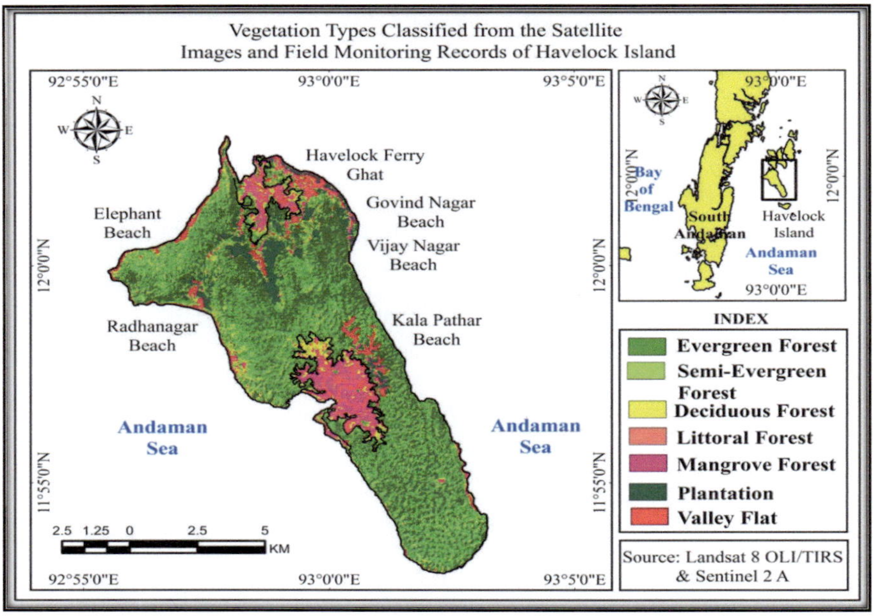

Fig. 2.38 Major forest types and their distribution in Havelock and Neil Island (2018)

Neil Island; and Bijoynagar, Radhanagar and Krishnanagar areas of Havelock Island in South Andaman.

2.2.8.2 Changing Forest Cover Areas

Total forest cover areas under total geographical areas of three different islands have been estimated and compared them with previous forest cover areas of 1986 in this section. Reduction of forest areas in comparison with 1986 indicates that a significant loss is responsible for spread of agriculture and settlement in remote corners of the islands. Tourism recreational infrastructure development along the coastal belt is also responsible for quick reduction in forest cover areas. Finally, it is also recorded that shoreline retreats and external sediment input into the shore platforms causing damages to the forest cover areas on the shore fringes of South Andaman by large-scale marine inundations (cyclones, tsunami incident, swell waves) in the previous decades (Table 2.13, Fig. 2.41).

Forest ecosystems and their stability regulate local hydrological cycle, recharge groundwater, maintain soil nutrients, and protect soil erosion on the hill slopes, supply forest products, influence wild life conservation and support environment-friendly tourism process in these islands. Isolated oceanic islands always try to adjust with changing environment by spreading their forest cover areas and forest types to

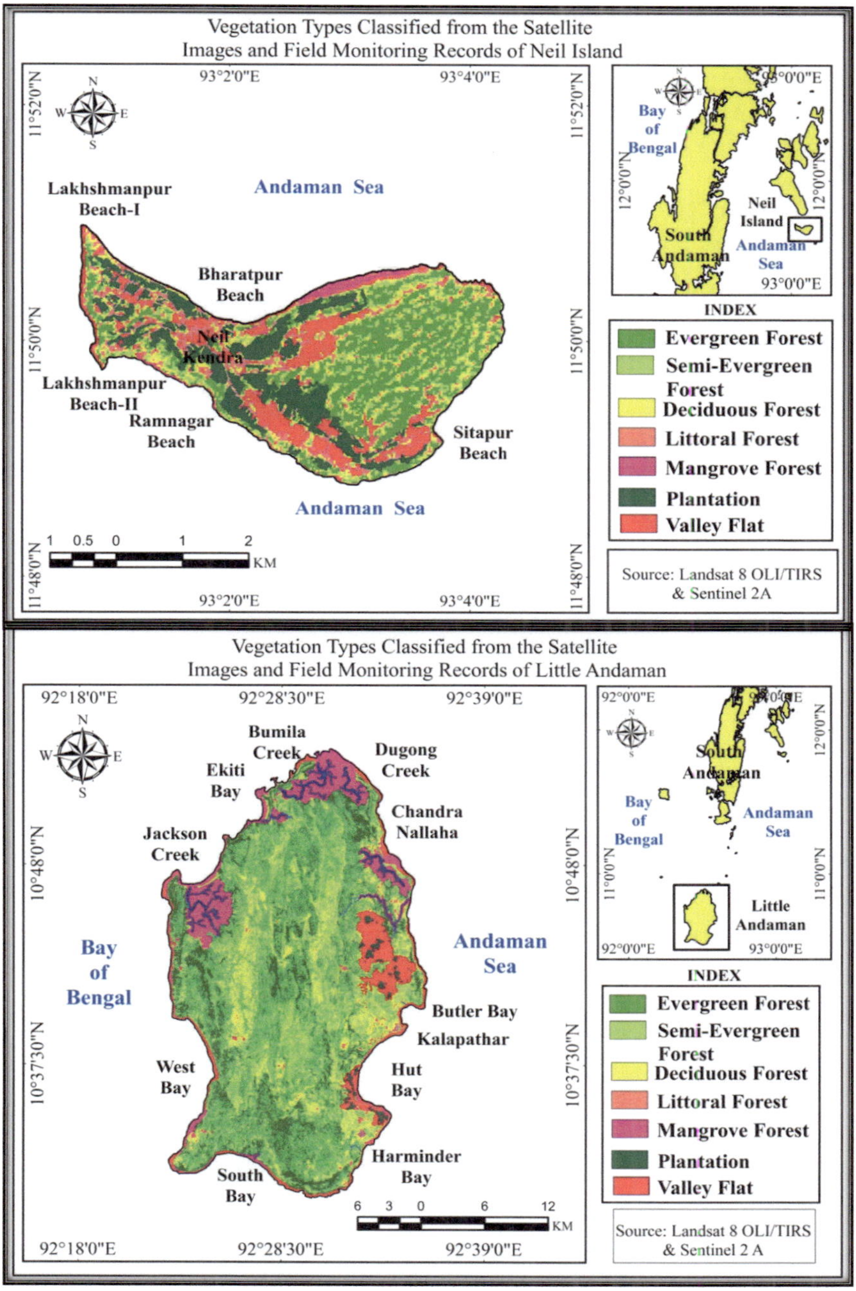

Fig. 2.39 Diversity of forest types and their distribution in Neil Island and Little Andaman

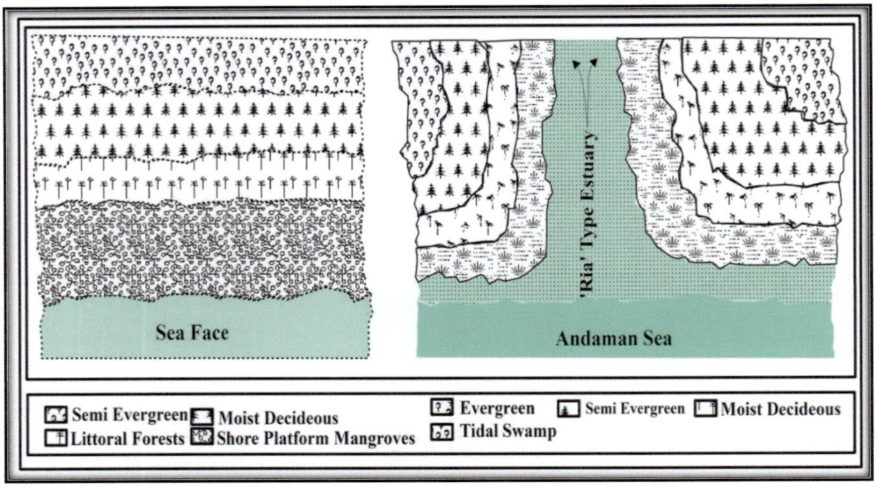

Fig. 2.40 Shore platform mangroves and estuarine fringe mangroves of Andaman Islands with their ecotone areas

Table 2.13 Changing forest cover areas in the islands

Islands	Total geographical area (Hectares)	Total area of forest (Hectares)	
		1986	2018
Little Andaman	73,157	70,358	61,027
Havelock Island	11,393	9903	8625
Neil Island	1890	676	563

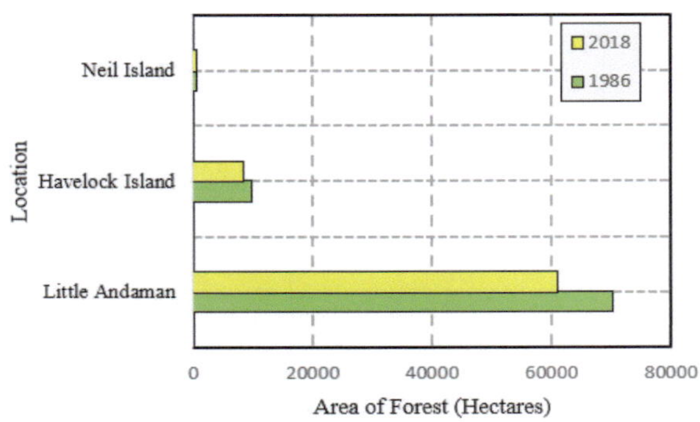

Fig. 2.41 Changing evergreen forests between 1986 and 2018

maintain the carrying capacity of island ecosystem. Overuse of forest resources by resident people and tourists may disturb the carrying capacity of Neil Island and Havelock Island in near future in South Andaman district. It is also observed that many types of marine litters (plastic bottles, plastic packets, glass bottles, straw materials, etc.,) are accumulated at the high tide level nearby the Neil Jetty centre and also in Lakshmanpur-II of the Neil Island and Havelock Island due to the tourism impacts at present (Table 2.13, Fig. 2.41).

Many trees are uprooted and usually fallen over the beaches or shore platforms due to land erosion, cliff recession and coastal landslides after the events of large-scale marine inundations and seismo-tectonic activities in Andaman Islands. Thus, logdrifts are common phenomenon along the shores (Fig. 2.42a) which can accumulate large number of wood logs in the sheltered embayments, bays and coves by long-shore transport. Field studies reveal that most of the dead trees are uprooted trees across the sea shores are fed by insects and gradually became decomposed and mixed with the soil as supply of nutrients in the oceanic environment. The dry leaf falls from the deciduous trees and semi-evergreen trees contributed dry biomass over the raised beach materials to add the ready nutrients after decomposition of leaves into the porous carbonate soils of these islands (Fig. 2.42a, b, c).

A narrow strip of littoral forest along the sandy shores or along the shore fringe reef terraces highlight the scenic beauty of the island shores in South Andaman. Diversified species from bushes, small trees and up to long trees with extensive branches provides shades and shelters for animal in the region. Major plant species of these category includes wild cashewnut trees, pandanus trees, sea mohul trees, wild jamun trees, Indian coral tree, bullet wood tree, the Alexandrian Laurel tree, horse-tail tree, tulip tree, etc., along these islands of South Andaman district (Fig. 2.42).

The leaves of littoral forest are thicker, oily, thorny, and belong to flowering plants with products of nuts and edible fruits along the littoral tracts can protect the low cliffs from the impacts of monsoon waves and swell waves for supporting the attractive coastal habitat in the coastal belts. Forest department has planted coconut trees, arecanut trees and casuarin trees along such sandy shores backed by reef terraces to introduce eco-tourism developments and also to manage the coastal environment in the region of oceanic islands (Fig. 2.42a, b, c).

2.2.8.3 Canopy Coverage with LAI Index

Canopy coverage becomes dense in the rainforest due to competition between the trees to get sunlight, and adjust with layered structure and extension of canopy area. Leaf area extension depends on chlorophyll content as photosynthetic pigment for the plant, and largely determines the photosynthetic capacity as well as the plant growth. In the tropical warm humid climate, strong evapotranspiration rate is enhanced with the availability of plant moisture supply to the photosynthetic process, which increase the chlorophyll content into the plant leafs. Evergreen trees extend their wide canopy coverage area in the forest with regular moisture supply. However, the tropical monsoon forests in the terraces and valley flats shed their leaves during

Fig. 2.42 **a** Buttressed trunks within the Andaman Moist Deciduous Forest in Havelock Island and Little Andaman; **b** Buttressed trunk of Jungly Dungy (Netajinagar, Little Andaman) and **c** Drifted wood log across the sea beach uprooted from the Coastal fringe forest

the period of reduced moisture supply in the rain less months in the islands. Thus, the mixed forest types of certain areas displayed fewer amounts of LAI values in the region (Fig. 2.24).

The Leaf Area Index (LAI) maps are prepared for the each island in the present section to show the temporal and spatial change of canopy characters of vegetations. The annual variation of rainfall, temperature and wind direction controls the supply of moistures into the forests. The annual diversity is grouped into pre-monsoon (January, February, March and April), monsoon (May, June, July, August) and post-monsoon (October, November, December) seasons for understanding the nature of LAI changes on the tropical maritime environment. The high values of LAI are observed during the month of May and during the month of March when significant amount of rainfall occurs in the Havelock Island. However, the LAI values vary during the post monsoon month (December) and pre-monsoon month (January) in the island tract. During this period, the island recorded lowest amount of rainfall. After all the Havelock Island exhibited significant low values of LAI in the central basin fringe area and northern valley flat areas (Figs. 2.43 and 2.44).

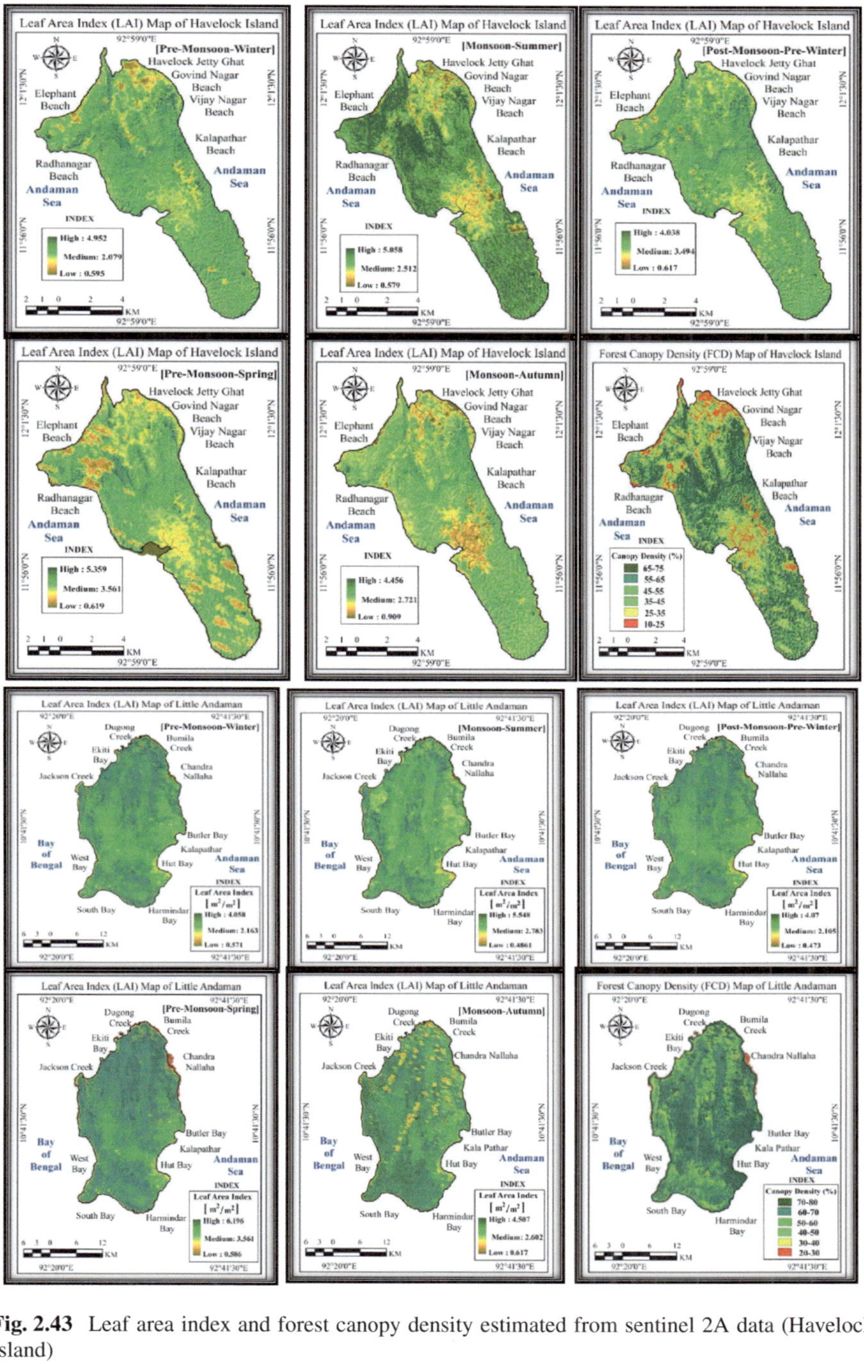

Fig. 2.43 Leaf area index and forest canopy density estimated from sentinel 2A data (Havelock Island)

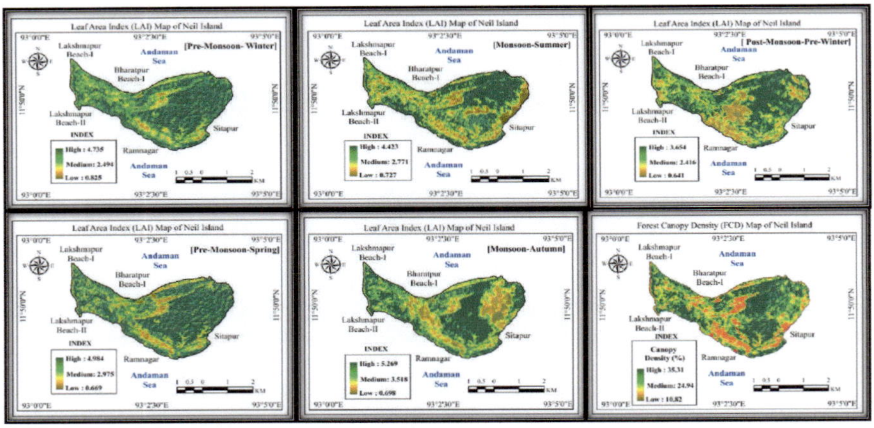

Fig. 2.44 FCD and LAI of Neil Island (estimated from satellite data, 2018)

Finally, the Forest Canopy Density (FCD) map of Havelock Island represents 65–75% canopy cover at the hill crests and 55–65% canopy cover at the piedmont slopes of the region. Other undulated terrains occupy the forest areas of moderate percentage FCD, ranging from 45 to 55% and 35 to 45% values (Figs. 2.43 and 2.44).

Little Andaman Island registered high values of LAI at pre-monsoon (March) and early monsoon (May) months, when they are showing as respectively. Such high values of LAI are responsible for occurrences of the high rainfall (3500–4000 mm per year) in the island (Figs. 2.45 and 2.46).

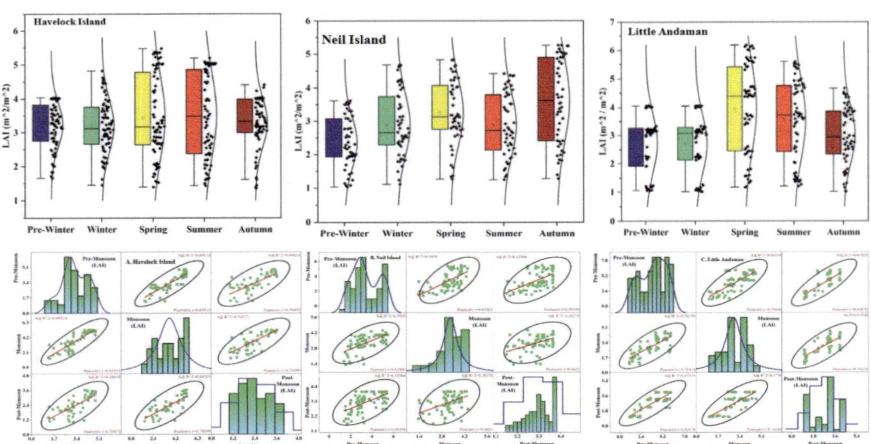

Fig. 2.45 Average LAI variation in the three islands estimated through Box-plot graphs and estimated Pearson correlation values for three seasons, showing high values for pre-monsoon and monsoon seasons

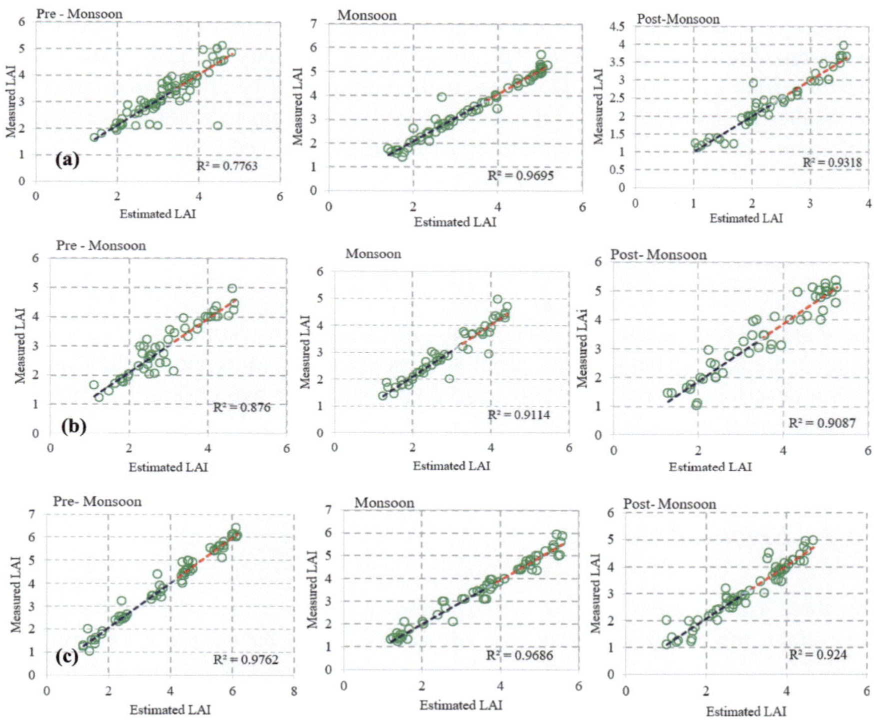

Fig. 2.46 Regression analysis between measured LAI and estimated LAI of all three islands **a** Havelock Island, **b** Neil Island and **c** Little Andaman

The LAI values are recorded at September, October and January month respectively in Little Andaman Island due to the occurrences of moderate amount of rainfall (3000 mm per month). Mainly the coastal fringes of Little Andaman represented lower amount of LAI in comparison with the island interiors.

The Forest Canopy Density (FCD) map of Little Andaman registered 70%–80% and 60%–70% canopy cover areas in the highlands interiors. However, the terraces of the island represented 50%–60% and 40%–50% canopy cover areas respectively. The high canopy density (>60%) of forest supports rich biodiversity, and for the reasons of high FCD the island provided quality forest resources for the aboriginal tribes (Onges and Nicobaries communities).

Neil Island is mostly modified for the development of tourism recreation infrastructures since 1985 in the South Andaman district. The eastern highland surface is only covered by dense forest. However, the shore fringe areas with western terrace land are moderately covered by forests. The LAI values are estimated for the island (relatively smaller in size) in late monsoon and pre-monsoon months when a sufficient amount of rainfall (> 2500 mm) occurs in the region. The LAI value shows only significant values in the post-monsoon month (December) in the island (Fig. 2.43).

The Forest Canopy Density (FCD) map of Neil Island showed only 35.31% in the eastern highland, 24.94% in the littoral fringe, and only 10.82% in the island interior terraces. Such low FCD values do not represent the healthier environment of the smaller island to support multiple activities like tourism-recreation, agriculture, settlements, pasturing, etc., over a long period (Figs. 2.45 and 2.46).

2.3 Evolution of Embayed Coasts

The shoreline configurations of South Andaman Islands are characterized by a series of embayments with bays, coves, promontories, ria-type shores and headlands. This section describes the nature of embayed coasts and their evolution for identifying the stages of development in different coastal sections.

2.3.1 Embayed Coasts

The three islands of South Andaman (Havelock, Neil and Little Andaman) of archipelago formation are considered for assessment of the evolution of embayed coast (Table 2.14, Figs. 2.47, 2.48 and 2.49). Shoreline types of the island configuration are classified and identified following the method of Strahler (1960) on the basis of SENTINEL 1A data and survey of India's (SOI) toposheets (Havelock Island: 87A/13, 86D/16, 86H/4 and 87 E/1, Neil Island: 87E/1 and Little Andaman: 87B/5, 87B/6, 87B/9, 87B/10) of the island group. Shoreline configurations influence over the waves and tidal regimes of the coastal system in the oceanic environment, and experienced the rate of energy input to modify the embayed shores through erosion, transportation and accretion processes.

There are ten shoreline types along the coastal sections of Neil Island on Andaman Sea. Among the total length of shoreline (25,188.09 m or 25.19 km), the length of headland beach occupies 5046.53 m or 5.05 km section and the length of the shoreline type of Bay Head Delta occupies 536.45 m or 5.36 km section in the island. It is also observed that four types of shoreline diversity such as bay side beach, headland beach, cliffed headland and cove jointly occupy 17,525.18 m or 17.53 km length of the island.

The shoreline types are categorized into 12 sections along the coastal parts of Havelock Island. The island has 86,439.35 m or 86.44 km length of shoreline and among them only lagoon fringe shoreline extends about 27,976.45 m or 27.98 km length and the cuspate bar shoreline type occupies 646.09 m or 6.46 km length of the coastal section. The major shoreline types are lagoon fringe shoreline, cuspate headland, bay head beach and bay side beach, and they jointly occupy 68,308.4 m or 68.30 km length (Fig. 2.49).

Little Andaman shoreline is categorized into 11 (eleven) types and the entire island has 203,562.74 m or 203.56 km length of shoreline in the South Andaman

Table 2.14 Types of configured shorelines of Havelock, Neil and Little Andaman Islands

Data base on shoreline configuration types (evolution of embayed coasts of Little Andaman, Havelock Island and Neil Island)				
Sl. no	Types of shoreline configurations	Shoreline configurations length (m)		
		Havelock Island	Neil Island	Little Andaman
1	CH	14,699.2805	3789.743643	14,188.44986
2	BHB	13,273.91107	1754.314112	26,131.79265
3	HB	4913.279387	5046.537421	26,850.51134
4	BSB	12,358.77321	4366.268645	26,876.71736
5	CS	916.452239	0	0
6	BMB	746.524438	1950.265845	6708.277002
7	CD	856.350789	0	0
8	CV	1100.458084	4322.613025	12,830.92109
9	CB	646.097718	843.512957	5604.420843
10	BHD	7303.187177	536.447419	24,611.18017
11	P	1648.583899	987.207896	65,730.23823
12	L	27,976.44286	1591.188651	5000.45933
13	I	0	0	8453.285961
Total length (m)		86,439.35	25,188.09	203,562.7382

Note CH—Cuspate Headland, BHB—Bay Head Beach, HB—Headland Beach, BSB—Bay Side Beach, CS—Cuspate Spit, BMB—Bay Mouth Bar, CD—Cuspate Delta, CV—Cove, CB—Cuspate Bar, BHD—Bay Head Delta, P—Promontories, L—Lagoon, I—Inlet

district. Among them bay side beach extends about 65,730.24 m or 65.73 km length of shoreline and the minimum length of 5000.46 m or 5 km shoreline is occupied by lagoon fringe shore in this island. Promontories, bay side beach, Bay Head Delta and Headland beach type shorelines extend jointly 144,068.66 m or 144.07 km length.

All three islands are now compared in terms of modification ratio as 12:4 (Havelock), 11:4 (Little Andaman), and 10:4 (Neil Island) in shoreline diversity and also compared in terms of total lengths versus length of major shoreline types occupied in the islands as 86:68 (Havelock), 204:144 (Little Andaman) and 25:17 (Neil). When the ratio of shoreline diversity is calculated we get the result as 3.0 (Havelock), 2.75 (Little) and 2.50 (Neil); but the ratio between total length and the length of major shoreline types exhibits 1.27 (H), 1.42 (L) and 1.48 (N). However, the standard deviation values and mean values of the shoreline types and shoreline length represent 2.75 ± 0.25 and 1.39 ± 0.11 respectively (Fig. 2.49).

Thus, the result shows that shoreline type is more significant than shoreline length. When the shoreline type is compared with the shoreline length for each island, we get the maximum value 1.48 for Neil Island, moderate value 1.42 for Little Andaman and minimum value 1.27 for Havelock Island.

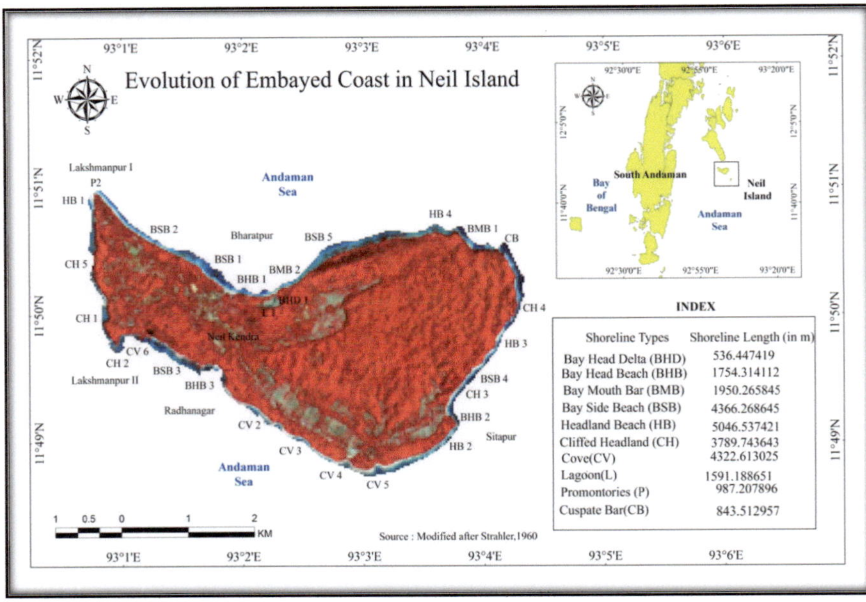

Fig. 2.47 Forms and types and evolution of embayed shore characters (Neil Island)

2.3.2 Stages of Development

On the basis of such values we can say that Neil Island is more modified by energy input and sediment movement, Little Andaman is relatively less modified due to less exposure, and Havelock island is least affected by energy input due to minimum exposure in comparison with the other islands of the region of Andaman sea and Bay of Bengal. Finally, we can say that Neil Island is in the old stage of shoreline modification, Little Andaman is in the matured stage of shoreline modification and Havelock Island is in the late youth stage of shoreline modification in the region (Fig. 2.50).

The Neil Island is located in the Andaman Sea, and most of the shorelines of the island are exposed to open marine swell waves, northeast monsoon drift and southwest monsoon drift associated with incoming waves of wind generated stress and cyclones.

When northeast monsoon winds start from the month of November to February the north and east of the island shoreline act as windward side, and during southwest monsoon winds southern and western sections of the shoreline act as windward sides particularly in the months of March to October. Thus, the shoreline modifications of Neil Island are experienced by erosion and sediment movement. However, the Little Andaman Island is located in the main island group over the accretionary prism and only exposed to the west coast swell waves and east coast swell waves of the Bay of Bengal and Andaman Sea.

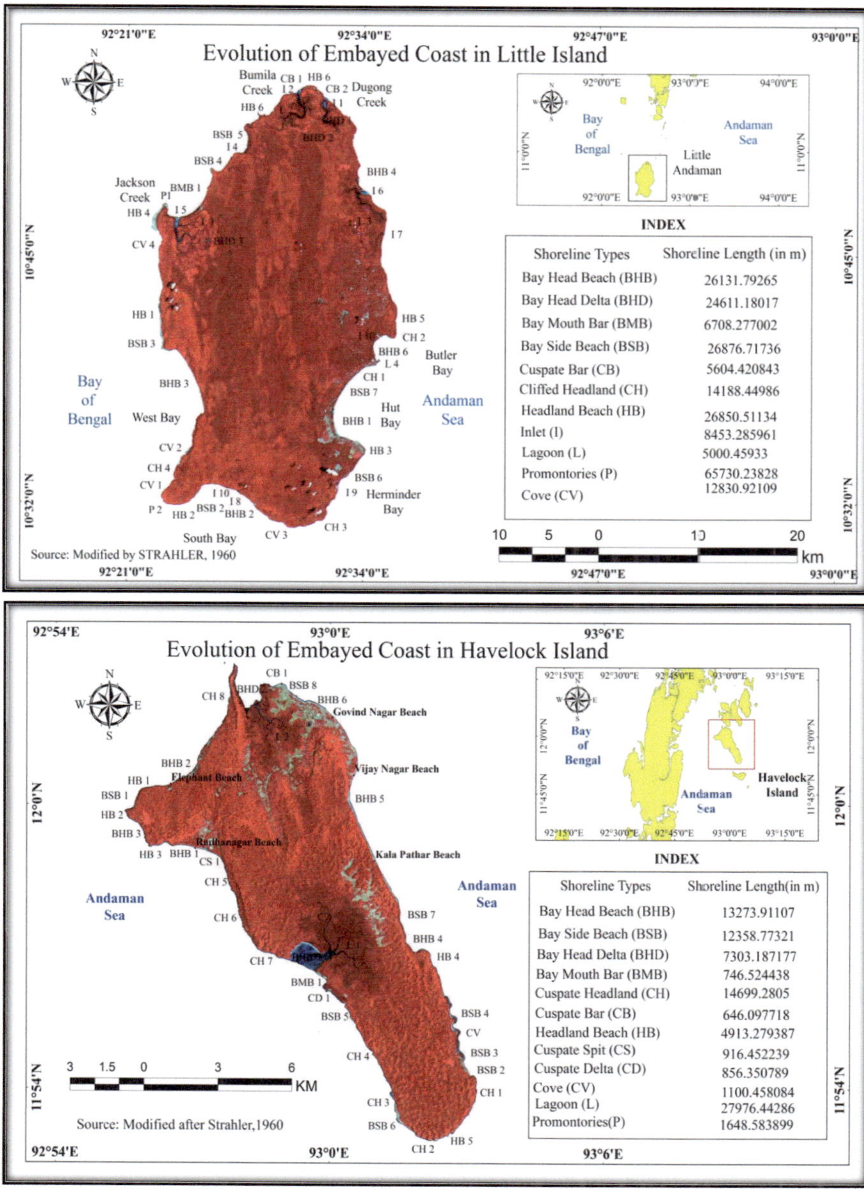

Fig. 2.48 Forms and types and evolution of embayed shore characters of Little Andaman and the diversified shoreline configuration types of Havelock Island

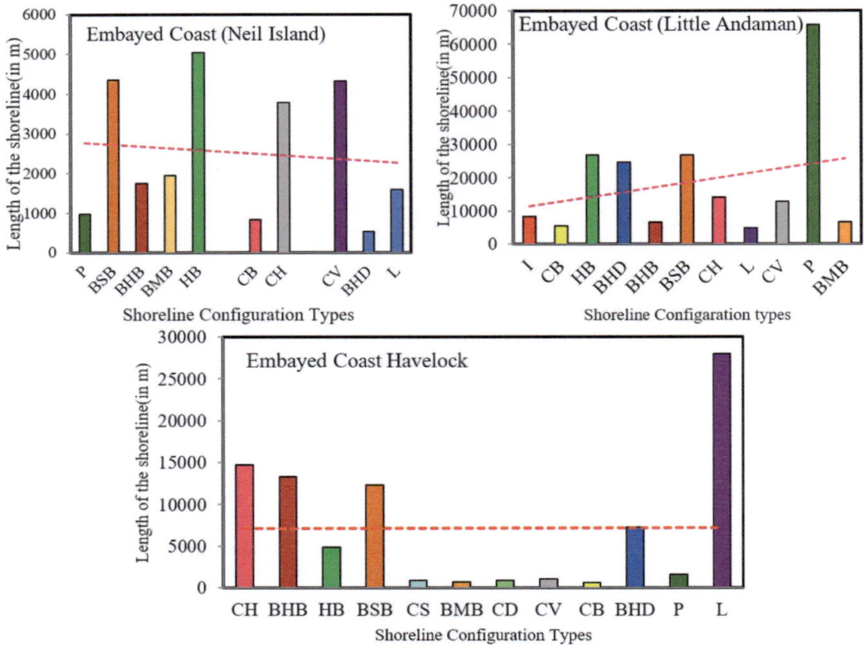

Fig. 2.49 Length of configuration types occupied by shoreline sections of the islands

The island is relatively large in comparison with the others and thickly vegetated from all sides in the regional setting. The southern part of the Little Andaman is fringed with 10° (ten degree) channel and thus the coastal parts of the region get affected by southwest monsoon winds. Carbonate shore platforms of the headlands, bay and promontories of the region are largely affected by erosion and sediment transport (Figs. 2.51 and 2.52, Tables 2.15, 2.16 and 2.17).

Havelock island of Archipelago group is protected from north and west by other islands of Ritchie's Archipelago and main group of Andaman. The eastern and south-eastern parts of Havelock Island are on the other hand exposed to the east coast swell waves. Majority of the shoreline types is located in the sheltered parts of island configurations and for which modifications are restricted and more or less controlled in many places of the regional setting (Table 2.17 and Fig. 2.53).

2.4 Coral Reef Coasts and Sediment Products

The Havelock Island, Neil Island and Little Andaman are coral fringed coasts in South Andaman district. The reef building corals of the region are dominated by cup corals, table corals and finger corals. So far as the data is concerned from MOEF

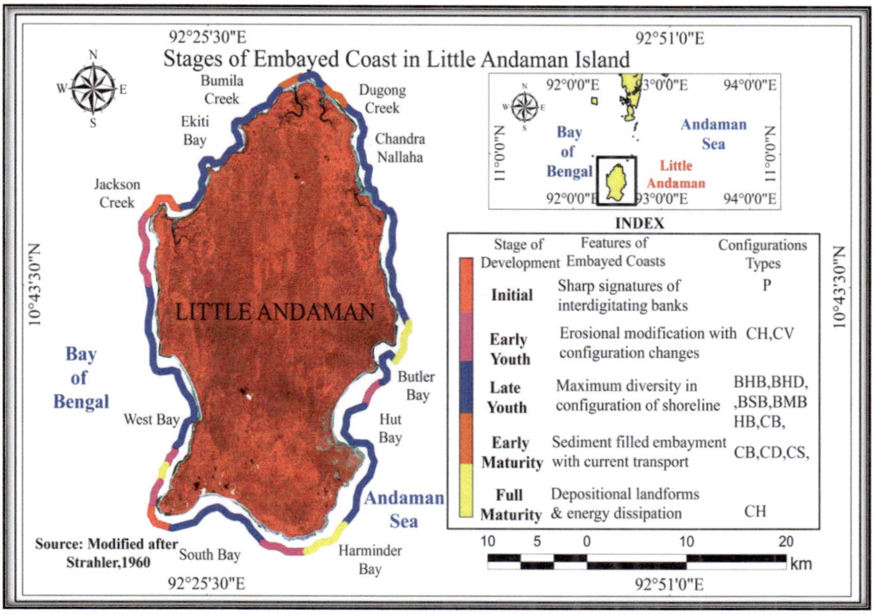

Fig. 2.50 Evolutionary stages of shoreline configuration types by the marine and coastal processes over the geological time

the total number of coral species recorded as 179 species belonging to 61 genera in Andaman and Nicobar Islands. The transparent sea water of the island fringes provided the condition for penetration of sunlight up to 50 mts depth and supported luxuriant growth of fringe corals in these islands. The zooxanthellae can photosynthesize from the light penetration and produces the supply of foods for coral polyps living within the skeleton of calcium carbonate deposited branches (Fig. 2.54) of reef morphology. During the event of the Great Earthquake (2004), the zooxanthellae has been reduced significantly in Havelock and Neil Island of the present study areas and 20% reef building corals were damaged by tsunami-generated sediment depositions (Fig. 2.54).

The Sen2Cor Model is prepared for all three islands individually with input of multiple data set of Sentinel 2A (Multispectral) Satellite image, Bathymetry data, and enhancement techniques and validation of data for final estimation of coral reef modelling in this part of the section.

However, the NOAA and GEBCO data are processed for generation of bathymetric contours of the Havelock and Neil Island, and Little Andaman Island. The depth contours are ranging from 10 to 600 mts for the Havelock and Neil and 6 to 1433 mts for Little Andaman Island in South Andaman Sea. Bathymetry contours are also showing their signatures of Jarawa Fault ridge (eastern margin of Little Andaman) and Eastern Diligent fault ridge (eastern margin of Havelock and Neil Island) on the sea beds (Figs. 2.55, 2.56 and 2.57). The coral reefs are located at the depths between

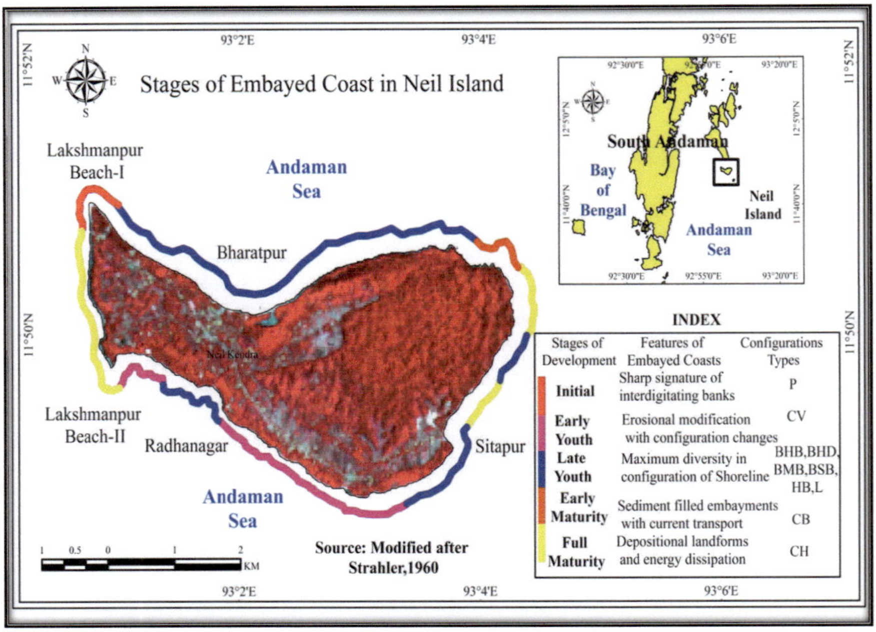

Fig. 2.51 Evolutionary stages of shoreline configuration types in the Neil Island by the marine and coastal processes over the geological time

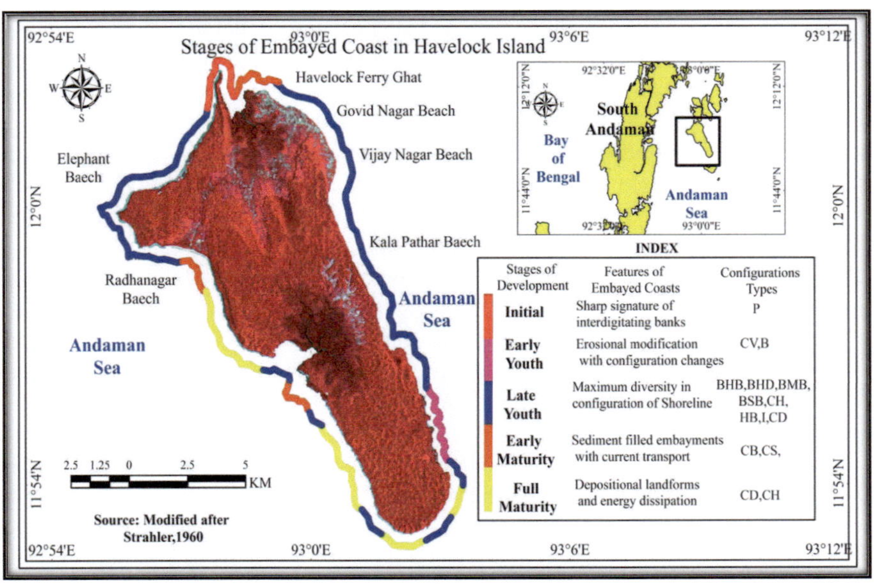

Fig. 2.52 Changing nature of shoreline configuration types through stages of development

Table 2.15 Multiple stages of embayed coasts evolved through the geological time scale along the shoreline sections of Little Andaman

Evolution stages and development of embayed Coast Little Andaman					
Sl. no	Stages of embayed coasts	Features of embayed coasts	Configurations types	Shoreline length (m)	Band color
1	Initial	Sharp signature of interdigitating banks	P	7614.36	Red
2	Early youth	Erosional modification with configuration changes	CV,CH	35644.05067	Pink
3	Late youth	Maximum diversity in configuration of Shoreline	BHB,BSB, HB, CB, BHD, BMB	144470.2015	Blue
4	Early maturity	Sediment filled embayments with current transport	CB, CD,CS	5840.065	Brown
5	Full maturity	Depositional landforms and energy dissipation	CH	9994.06	Yellow
Shoreline length of Little Andaman				203562.7382	

Table 2.16 Various features of embayed coasts and their stages of development in Neil Island

Evolution stages and development of embayed Coast Neil Island					
Sl. no	Stages of embayed coasts	Features of embayed coasts	Configurations types	Shoreline length (m)	Band color
1	Initial	Sharp signature of interdigitating banks	P	1177.21	Red
2	Early youth	Erosional modification with configuration changes	CV	4781.24	Pink
3	Late youth	Maximum diversity in configuration of Shoreline	BHB,BHD,BMB,BSB, HB,L	13365.84	Blue
4	Early maturity	Sediment filled embayments with current transport	CB	1017.699614	Brown
5	Full maturity	Depositional landforms and energy dissipation	CH	4846.11	Yellow
Shoreline length of Neil Island				25188.09961	

Table 2.17 Evolutionary stages of shoreline configuration types of Havelock Island

Evolution stages and development of embayed Coast of Havelock Island					
Sl. no	Stages of embayed coasts	Features of embayed coasts	Configurations types	Shoreline length (m)	Band color
1	Initial	Sharp signature of interdigitating banks	P	0	Red
2	Early youth	Erosional modification with configuration changes	CV	10289.05	Pink
3	Late youth	Maximum diversity in configuration of Shoreline	BHB,BHD,BMB,BSB, HB,L	11034.37	Blue
4	Early maturity	Sediment filled embayments with current transport	CB, CS	44886.14137	Brown
5	Full maturity	Depositional landforms and energy dissipation	CD, CH	20229.78	Yellow
Total shoreline length of Havelock Island				86439.34137	

10 and 50 mts in Havelock and Neil Islands and at the depths between 20 and 50 mts in Little Andaman Island along the seaward slopes of Andaman ridges.

The coral reef zonation mapping of Havelock Island represents algal ridge with coral reefs, rock coral algae with living reefs, coral rubbles with patch reefs, living coral reefs, sea grasses colony with coral rocks and rubbles reef with living corals and dead corals.

However, the Neil Island exhibited eight micro zones of coral reefs like all the above zones with a zone of corals with algae rubbles bed. The estimation also

Fig. 2.53 Cove beach in Sitapur section (Neil) and the rock ribs of the shore platform fringed with the cliff in Radhanagar west section (Havelock)

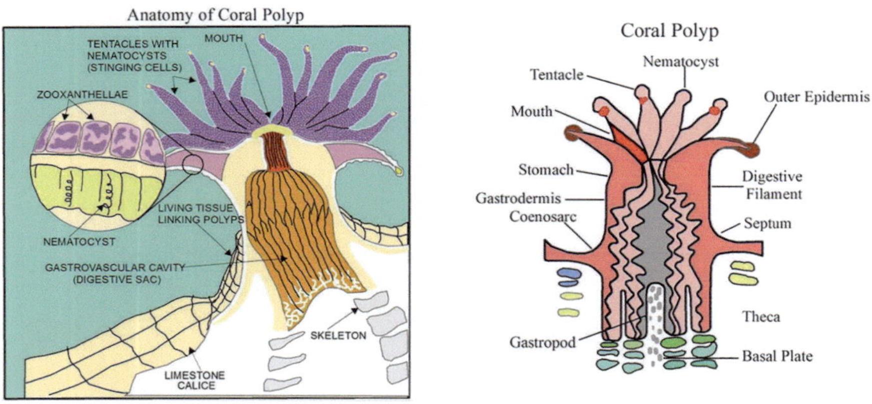

Fig. 2.54 Anatomy of coral polyp with the reciprocal relationships between zooxanthellae

shows that Neil Island support more luxuriant growth and expansion of coral reefs than Havelock Island (Figs. 2.56 and 2.57). During the event of tsunami (2004) in Andaman Sea, the Havelock Island reef habitat was more damaged than Neil Island. Little Andaman on the other hand also exhibited the similar types of eight coral reef zones and more luxuriant growth on the seaward front of the Bay of Bengal than the eastern sea front of Andaman Sea. Some reef areas of Little Andaman are also affected by upliftment due to the event of seismo-tectonic activities (2004) in the region.

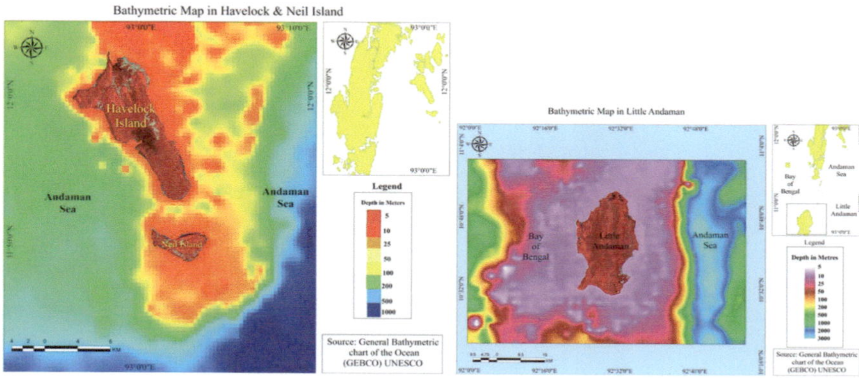

Fig. 2.55 GEBCO data represents bathymetric depth variations of the three islands (2018)

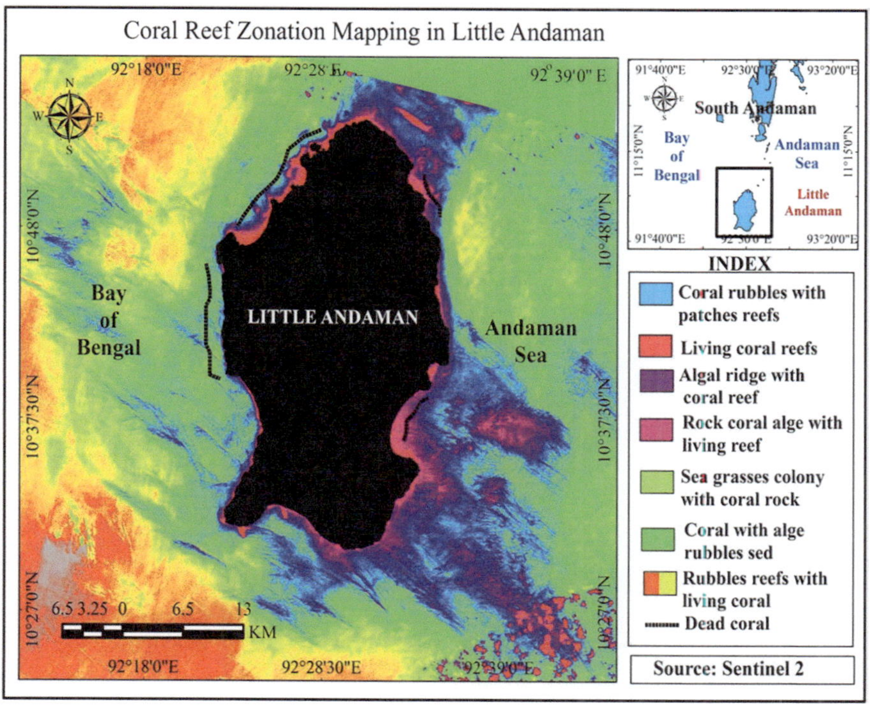

Fig. 2.56 Coral reef zonation map of Little Andaman based on Sen2Cor tool

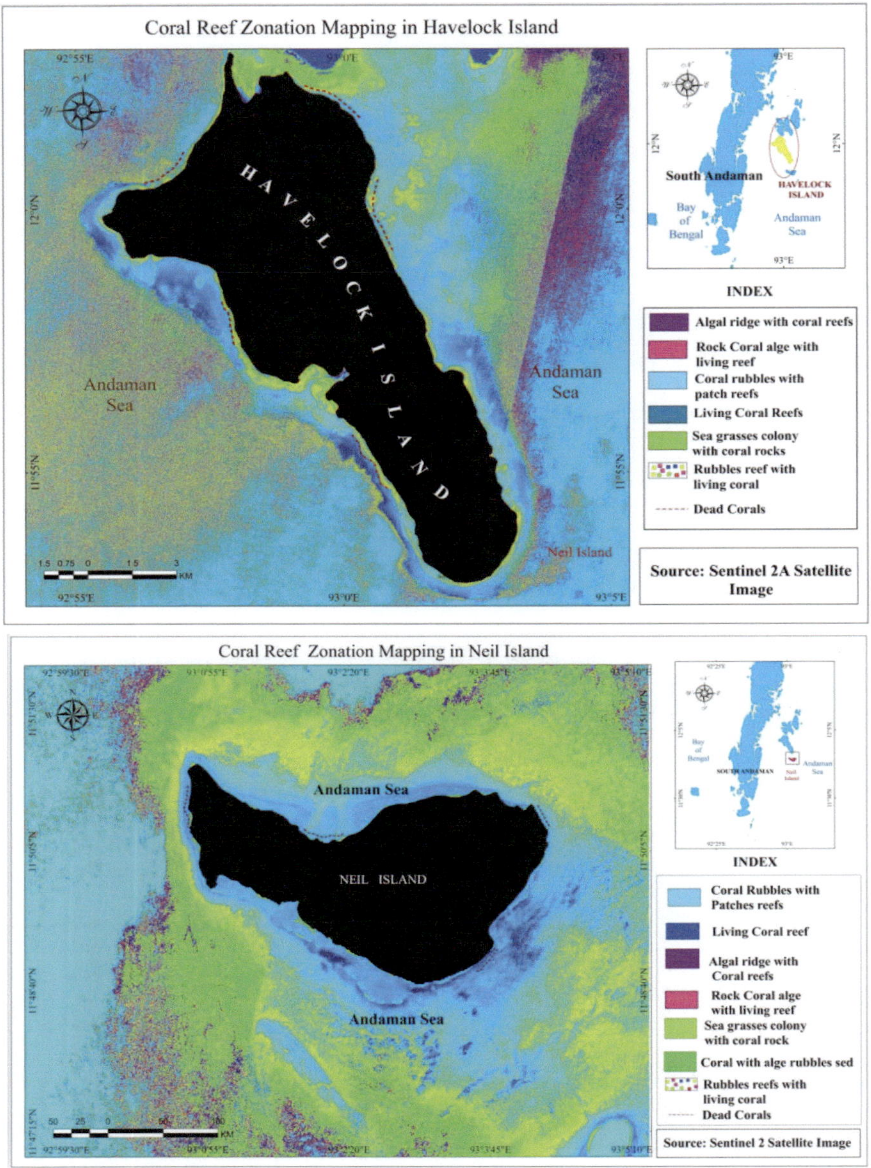

Fig. 2.57 Coral reef habitats mapping of Havelock and Neil Island using Sentinel 2A (Multispectral) data

2.4.1 Coral Reef Habitat

The fringing coral reef zones are identified through **Sentinel-2 data** of South Andaman Islands in the present study. Coral reefs by their species richness and physiographic location are grouped into: shallow reefs with algal ridges, reef front coral banks, terraced bank fringing reefs and attached reefs on the descending slope of the outer banks. Among them, the reef front coral banks are dominated by *Acropora, Porites and Montipora* species at the depth of 10 m bathymetric contour in Neil and Havelock Island.

Corals were bleached in the shallow reef flats with reduction of algal mats. However, the better reef growths are available at 15–20 mts bathymetric contours along the shore fringes of Little Andaman Island. The dead coral rubbles and debris are distributed on the landward margins of shallow reef flat zone, and other dislodged dead coral debris were transported seaward on the descending slopes of outer banks in the region after tsunami event (2004). However, the coral rubbles and sands were also moved onto the shore platforms by bulk transport during the events of marine inundations. The patch reefs of the lower intertidal shore platforms were blanketed by sediments during tsunami event in all three islands. Currently, they are re-emerged in the shallow platform surface dominated by location of frequent rock pools after wave actions over the years (Fig. 2.58). Tourism and recreational activities in the coral reef environment need to be restricted in a specific season to avoid the negative impacts. Marine protected areas of certain sections of the island shores are also needed to reduce the stressors in the coral reef environment at present (Fig. 2.58).

Fig. 2.58 Coastal Karren features and Taffoni marks on the shore platforms adjacent of the shore cliff and degraded patch reef by sedimentation in the rock pools

2.5 Shore Platform Surface Morphometry

The shore cliffs of Little Andaman, Havelock, and Neil islands (Figs. 2.59, 2.60, 2.61 and 2.62, Table 2.18) are fronted by various kinds of shore platforms. Near horizontal to sub-horizontal and gently sloping platforms of intertidal and sub-tidal environment are exposed to wave attacks, water layer weathering, salt weathering and subaerial weathering processes. The carbonate rocks of the shore platforms are dissected and wave-abraded on the windward sides of the above islands. In many cases the marine denudation and weathering processes have led towards shore planation in the tropical coast, and others are depositional in the sheltered coast with sediment cays in the region of Andaman Sea and Bay of Bengal.

Various kinds of shore platform includes as seaward sloping shore platform, high-tide shore platform, low-tide shore platform, and plunging cliffs with no platform on the shore (Bird, 2008).

Shore platforms are the result of combined effects of soft rocks, hard rocks, sea level changes, wave erosion rate, weathering rate, tectonics and humid climate in the tropical coasts. As the tidal range is only 1.5 mts and significant wave heights range from 2.0 to 3.0 mts on the shore fringe open marine environment, the wave abrasion and wave quarrying dominate over the weathering processes in the region. However,

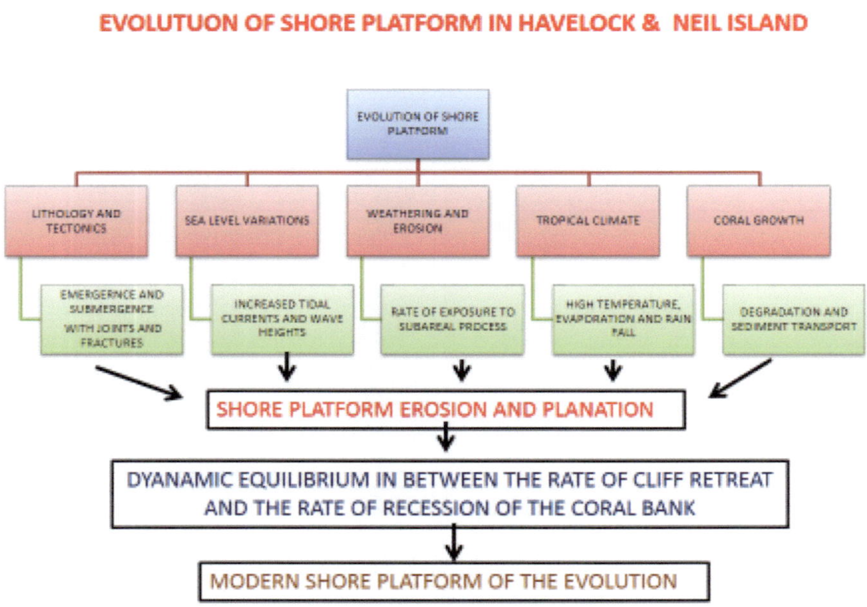

Fig. 2.59 Evolution of shore platform in Havelock and Neil Island

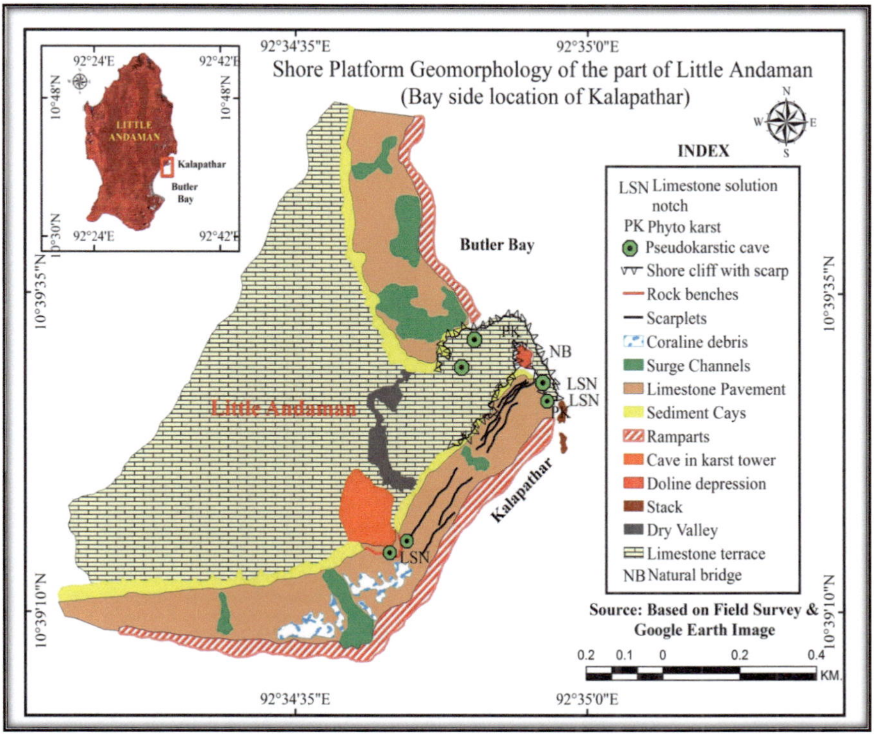

Fig. 2.60 Diversified shore platform morphology of Little Andaman

the signatures of wave abrasion, weathering with wetting and drying processes, solution processes, bio-erosion and induration on the shore reveal the diversity in the shore platform morphometry in the islands (Figs. 2.62 and 2.63).

The extent of various kinds of shore platforms is identified in three different islands on the basis of rigorous field work along the shorelines and identification of satellite image-based features of the shore fringe areas (Fig. 2.64).

2.6 Geological Outcrops and Lithology

Sediments of accretionary wedge at the convergent margin between the Indian Ocean Plate and Eurasian Plate have made the foundation of a long North–South trending ridge of Andaman Group of Islands on the western margin of Andaman Sea. This is a high relief fold thrust belt of sediment-dominated orogeny with accretionary wedge which formed along the obliquely subducting eastern margin of Indo-Australian

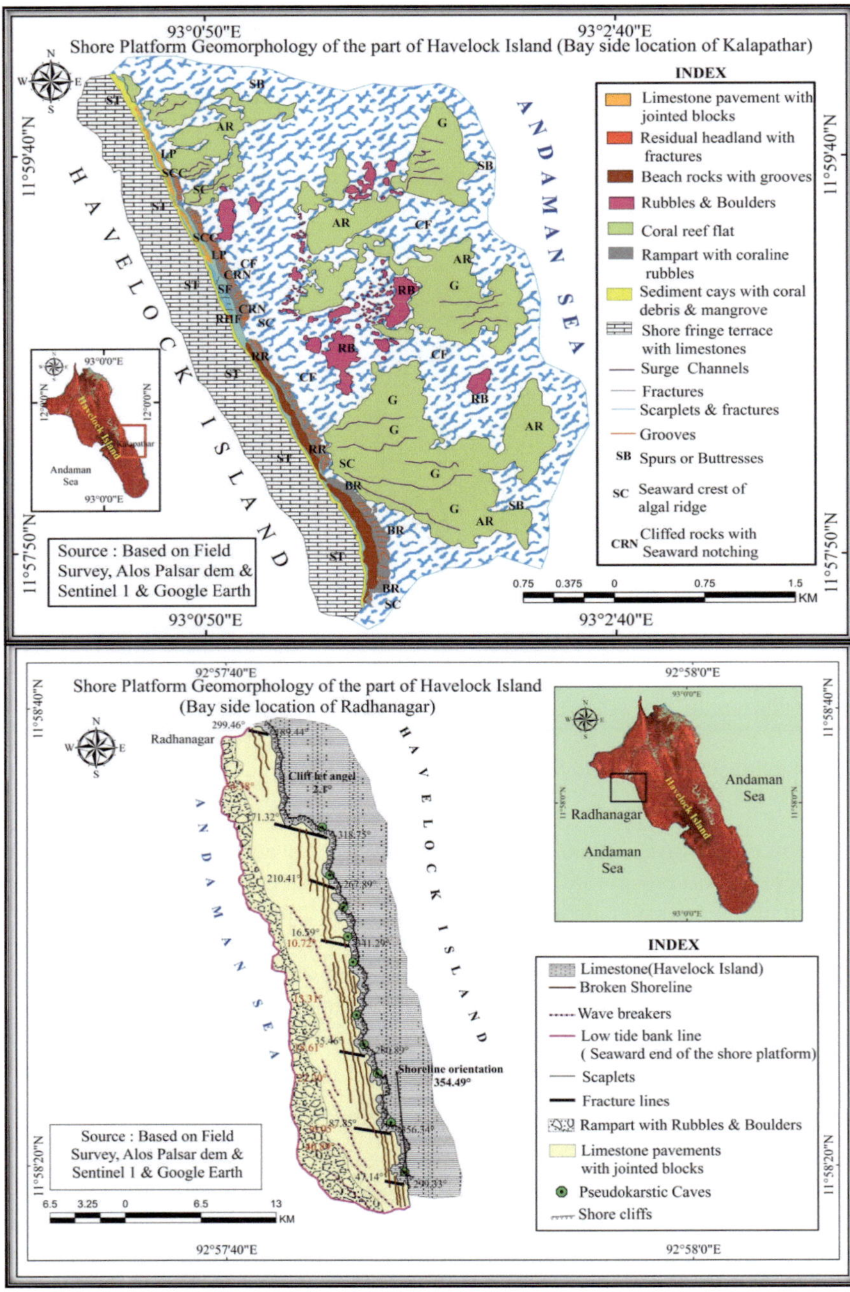

Fig. 2.61 East coast and west coast shore platform diversity in Havelock Island

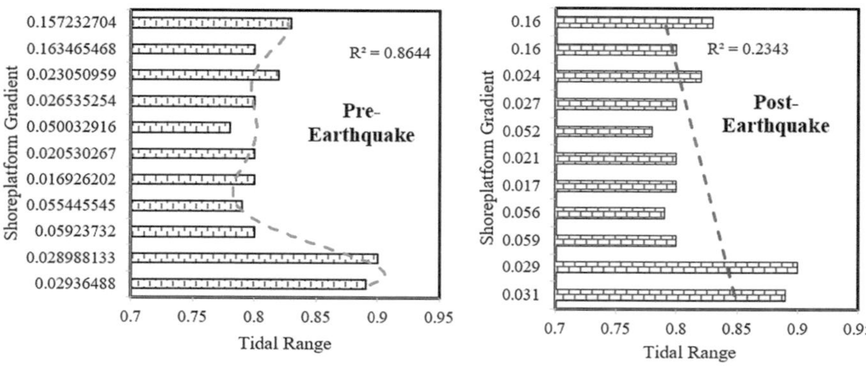

Fig. 2.62 Shore platform gradient showing the pre-earthquake and post-earthquake effects

oceanic lithospheric plate (Bandyopadhyay and Carter, 2017). However, stratigraphically the islands are made up of Pre-Cretaceous meta-sedimentary rocks, Upper-Cretaceous ophiolites and Palaeogene-Neogene sedimentary formations under active tectonic settings in the oceanic environment. The islands expose cretaceous rocks in several places and they are also intruded by Peridotites and serpentines in Upper Cretaceous period. Radiolarian cherts associated with ultramafic intrusives of late cretaceous age are found on the eastern margin of the islands. The Port Blair series of Eocene formation with the beginning of Tertiary contains Nummulites and other marine fossils. The overlying Greywacke series of unfossiliferous beds are formed in Upper Eocene to Upper Oligocene, and finally the Archipelago series with arenaceous sandstones containing Aquitanian fossils are lying on the top in the region of South Andaman. The unconformity is produced above the Archipelago series due to strong earth movements related with mountain building activities in the Miocene period. Series of White Shale has appeared above the unconformity and they belong to the age of Pliocene and little amount of Miocene Formation (Krishnan, 1982, Sharma and Srinivasan, 2007, Singh and Vimal, 1976). However, the youngest formations of Pleistocene sands, clays and coral banks are lying above the white shale series of Andaman Islands (Table 2.19).

The structure of Andaman-Nicobar ridge is described as accretionary prism of the subduction zone by Curray (2005), which morphologically represents from west to east as subduction zone, outer arc (Andaman-Nicobar Ridge), fore arc basin (west basin, invisible bank), volcanic arc (Barren, Narcondum) and the back arc basin (central Andaman basin, East Basin). The eastward dipping fault slices and folds are separated from fore arc basin by two significant normal faults such as Diligent Fault and Eastern Margin Fault in the regional setting. Structurally, the study areas of Havelock and Neil Islands are located on the fore arc basin and Little Andaman Island is located on the outer arc basin (Fig. 2.65).

The general geological succession of the Andaman and Nicobar Group of islands can be tabulated in this part of the work following the works of Karunakaran et al. 1964; Ray, 1982; and Vohra et al. 1989 to identify the geological characters and

Table 2.18 The differences in shore platform gradient of pre-earthquake and post-earthquake positions in Neil Island

Shore platform gradient of pre-earthquake and post-earthquake(26th December,2004) in Neil Island, South Andaman

Sl. no	Section point	Latitude	Longitude	Elevation (m)	Distance (m)	Tidal range (m)	Shore platform gradient (pre-earthquake)	Shore platform gradient (post-earthquake)
1	Neil Kendra I	11°49′ 51.28″ N	93° 1′ 10.17″ E	3.5	119.19	0.89	0.031	0.029
2	Neil Kendra II	11°49′ 38.86″ N	93° 1′ 36.53″ E	3.2	110.39	0.9	0.029	0.029
3	Ramnagar	11°49′ 10.84″ N	93° 2′ 14.56″ E	3.2	54.02	0.8	0.072	0.059
4	Atreya Mangrove	11°48′ 53.19″ N	93° 2′ 51.78″ E	2.8	50.5	0.79	0.061	0.056
5	Govindra Nagar	11°50′ 28.63″ N	93° 3′ 4.00″ E	4	236.32	0.8	0.021	0.017
6	Bharatpur I	11°50′ 11.41″ N	93° 2′ 19.80″ E	3.5	170.48	0.8	0.031	0.021
7	Bharatpur II	11°50′ 21.19″ N	93° 2′ 37.56″ E	3.8	75.95	0.78	0.027	0.051
8	Jetty Point I	11°50′ 16.72″ N	93° 1′ 45.13″ E	3.5	131.9	0.8	0.037	0.0267
9	Jetty Point II	11°50′ 24.17″ N	93° 1′ 33.90″ E	2.8	121.47	0.82	0.052	0.024
10	Lakshmanpur I	11°50′ 42.35″ N	93° 0′ 46.38″ E	4	24.47	0.8	0.341	0.163
11	Lakshmanpur I	11°50′ 32.24″ N	93° 0′ 47.21″ E	5	31.8	0.83	0.434	0.157

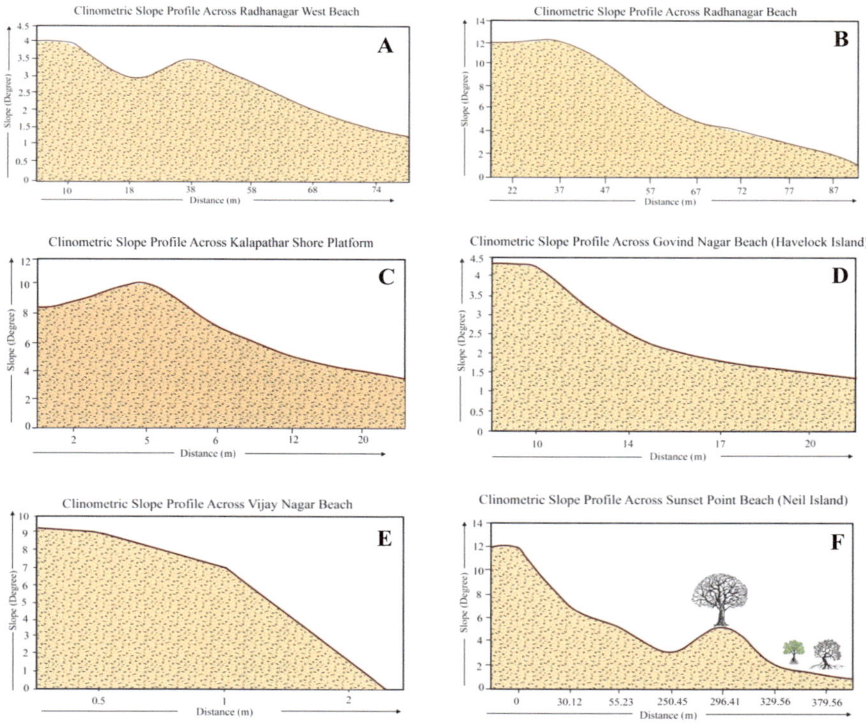

Fig. 2.63 Clinometric slope profile along the shore platforms of different sites in Havelock and Neil Island

Fig. 2.64 Measuring the shore platform gradient using the clinometer and scale

Table 2.19 Various geological formations with stratigraphic succession in Andaman Islands (after Chatterjee, 1967)

Formation	Stratigraphic succession of the Andaman Islands	
Recent formations	Swampy mud flats, coral rags and raised beaches	Recent and Pleistocene
Archipelago group formations	*Unconformity*	
	Long clay formation Guitar limestone formation	Pliocene-Miocene
	Unconformity	
	Round chalk and silt formation Strait sandstone formation	Pliocene-Miocene
	Unconformity	
Port blair formations	Port blair formation Greywacke, shales, siltstones and sandstones	(?) Oligocene-Late Eocene
	Unconformity	
	Igneous Series Ultrabasic intrusive mainly pyroxinites largely altered to serpentinites and extrusives	Palaeocene
	Igneous Contact	
Baratang formations	Baratang formation Shales, Sandstone, Limestones, Greywacke and Conglomerates	Palaeocene-Late Cretaceous
	Unconformity	
Pre-tertiary formations	Older sedimentaries Cherts, jaspers, quartizites and limestones	Mesozoic (extending into M. Cretaceous?)
Base not Exposed		

related lithological settings of the Havelock Island, Neil Island and Little Andaman Island in South Andaman (Fig. 2.65).

Chronologically, the geological ages of Precambrian rocks, Middle to Lower Eocene rocks, Oligocene to Upper Eocene rocks, Pliocene to Miocene rocks, and Holocene to Pleistocene rocks are represented by different exposure sites of the island groups. Several geological formations of Andaman and Nicobar groups of islands also grouped as older meta sediments, Ophiolites complex, Mithakari group, Port Blair formation, Archipelago and Pleistocene-Holocene materials, and they are significantly separated by unconformities of erosional contact and tectonic contact in the region between Andaman Sea and Bay of Bengal. The lithofacies from older to younger formations include as highly deformed quartzite's and hornblende schist at the basement; pillow lava, diorite, cumulates and peridotites as overlying sequence;

Fig. 2.65 Measuring the dip and strike of the rock strata using the Brunton Compass

sandstone, grit, conglomerate, cherty limestone and black carbonaceous shale overlying as shallow water and trench fill deposits; Andaman flysch sediments of alternate sandstone, siltstone, grit and conglomerates as overlying sequence; limestone, sandstone, chalk, white clay stone and conglomerate overlying as archipelago formation; and the top layer constitute recent carbonate deposits Pleistocene lava, ash and pyroclastics, mangrove clays, coral reefs and beach sands with raised beach features in the oceanic islands (Fig. 2.65). The islands of Richie's Archipelago are located on the forearc basin and constitute the Neogene sediments in a sequence of succession in Andaman Sea. Among the small group of islands, Havelock Island and Neil Island are considered for the present study in this regional setting of South Andaman district. Some of the researchers (Daneshian et al., 2007) reported about the occurrences of a late-early middle Miocene palynofloral assemblage from the chalk sediments and mudstones from Lithostratigraphic sections of the Inglis formation of Richie's Archipelago. However, the Neogene rocks of Archipelago group have been divided into seven formations from the Lithostratigraphic sequences by Srinivasan and Azmi (1976) and among them the 'Inglis Formation' was deposited at middle and lower bathyal depths. The 63 m thick sequence of sedimentary formation has been formed alternately with the layer of siltstone or mudstone in the stratigraphy (southwest coast section) of Inglis Island. This formation is very closely situated nearby the Havelock Island and Neil Island of the same island group of the Ritchie's Archipelago.

2.6.1 *Havelock Island*

Various cliff sections of the Havelock Island near Radhanagar east, Krishnanagar, promontory of Elephant beach, Kalapathar, etc. points indicated the occurrences of white clay stone, chalk, siltstone, mudstone, sandstone, shaley limestone, coralline limestone and conglomerates at the lithostratigraphy. Hills are mostly dominated by white clay stone and siltstone and densely covered by luxuriant growth of evergreen and semi-evergreen forests. Coral rags, beach sands, mangrove clays, carbonate sediments, coral reefs and beach rocks are overlying the stratigraphic sequences of the island system.

The Pleistocene massive limestones with graded sandstones occupy the basement areas of Havelock Island (Fig. 2.66). They are unconformably separated by overlying beds of shales, calcareous sandstones, mudstones, white clay stones, chalky limestones, black shales, sheety limestones and marls in the archipelago formations. Various exposure sections of the rocks beds along the western shore cliffs of Radhanagar, southeast and northwest shore fringes and Kalapathar represent significant structural deformations like faulting, tilting, jointing, upliftments and rupturing. Cave dissolution process is very active along the joints of the soluble rocks beds on the shore cliffs. The patterns of the strike and dip of the deformed rocks beds are measured in the cliff sections during field works in the island points. The strike of the rocks beds is aligned from SSE to NNW direction in the island and the beds are dipping towards east at an angle between 27 °C and 35 °C, and they have produced active cliff wall towards the sea face fringed with wave-abraded shore platforms, caves and notches are prominent in the re-entrants of the cliff line. Exposures of the older rock beds on the shore platforms have produced clifflets and covered by the apron of rock boulders (shales, sandstones, limestones, coral rags and fossil coral debris) in the region.

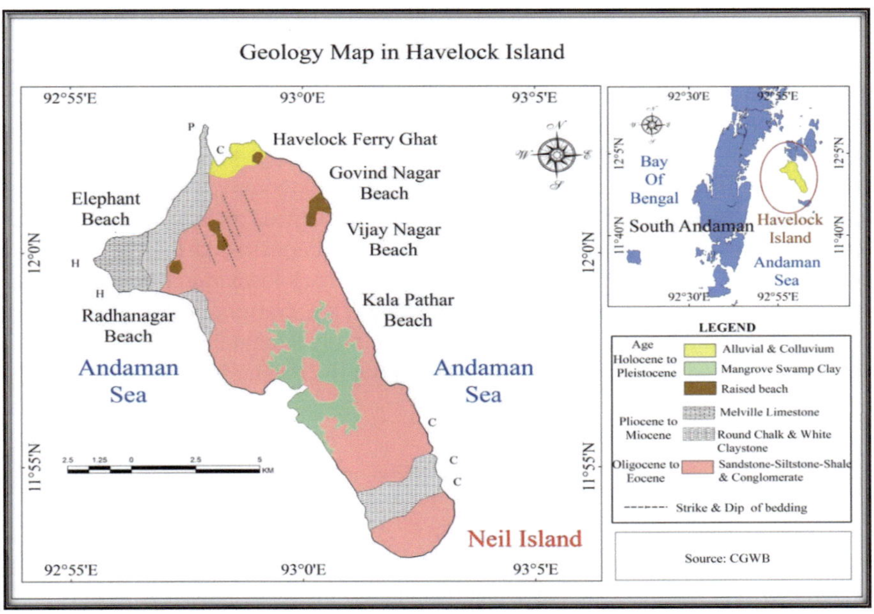

Fig. 2.66 General geology of the Havelock Island

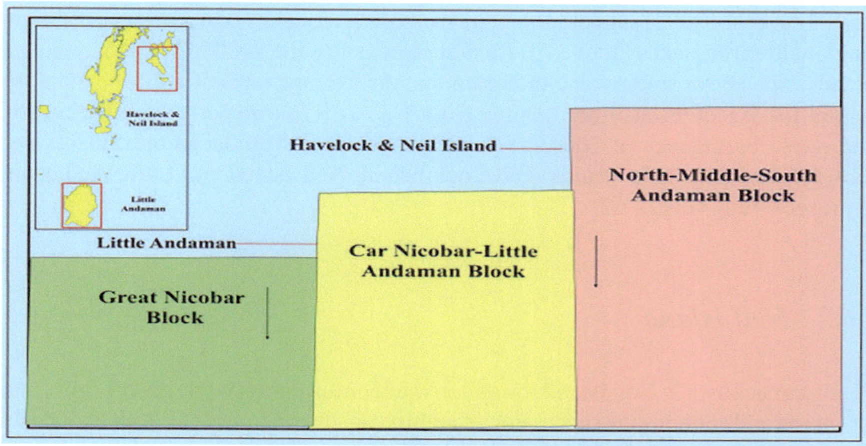

Fig. 2.67 Block-wise structural upliftment after the occurrence of great earthquake (2004) in Andaman and Nicobar Islands

The long cliff wall of NNW promontory shows the lower most white clay stone is overlain by successive layers of chalky limestone and shale, and as the rock beds are highly jointed so a series of fissure caves have been developed in the cliff section. Direct exposures of the rock beds are also absorbed along the shore face of a subdued headland in Kalapathar section. The seaward dipping rock beds include as alternate layers of massive sandstone shale and mudstone. The strike line of rock beds is directed towards NNW parallel to the shoreline and highly jointed and fractured. Caving and notching are common features on the east facing sea cliffs of the Havelock Island near Kalapathar section. The high cliffs with chalky limestone, white claystone and laminated shales are visible on the seashores of extreme south of the Havelock Island (Fig. 2.67).

The typical 'hogback' like south-facing hill fronts represents such types of stratigraphic section in this part of the island in Andaman sea. Similar stratigraphic sections are also deserved in the island interiors hills in southeastern part and north-central part of the island. The raised beach features and reef terraces are constituted by carbonate clays thick deposits of sands, coral debris and marine shales with debris of fossil corals cemented by carbonate materials and other bioclasts at different elevations. The shore platforms with wide exposure sites of Vijaynagar, Havelock jetty points, and coastal embayments of Havelock north, Elephant beach, Radhanagar and Havelock south support mangrove clays, carbonate clays, tidal sands and muds of younger sediments. Area's mangrove forests are thickly distributed in the region of carbonate settings of Havelock Island.

Ample of morphologic evidences are there in the island units in support of active seismicity and neo-tectonism. The radio carbon age data of coastal fossil coral terraces of number of islands in Andaman and Nicobar group (Awasthi et al., 2013;

Rajendran et al., 2008) suggest about the increased frequency of occurrences of high-magnitude earthquakes (above 7.0 Richter scale) since the 9000 YBP in the younger islands. The above studies also indicated that the average rates of upliftments of the coastal fossil reef terraces are ranging from 1.33 to 1.76 mm per year in the region. The recent earthquake of 2004 (>9.0 Richter scale) has produced upliftments and subsidences in all three islands (Havelock Island, Neil Island and Little Andaman) of present study (Fig. 2.67).

2.6.2 *Neil Island*

Stratigraphically, the Neil Island formation is unconformably overlying the Port Blair formation and contains the succession of archipelago formation along with Holocene-Pleistocene sequences. Number of coastal cliff sections at Neil west, Lakshmanpur-2 and Sitapur point represent clear stratigraphic sequences from Miocene to Holocene ages. The Neil west section represents highly fossiliferous layers of sheety limestone, massive limestone, calcareous sandstone, coralline limestone, shale and conglomerates. Various types of fossil corals, fossil bivalves and marine shells and other marine animal fossils are visible from layers of stratigraphic sequences in this part of the island. The layers are jointed, fractured and tilted by seismic events and ultimately separated into blocks over the subsided shore platform of intertidal region. Elevation of the cliff section was situated on 10 m surface above sea level in the form of significant layered structures. The broken layers and boulder size rock debris are lying over the foot slope or over the toe slope of fossil reef terrace on the western margin of Neil Island (Fig. 2.68).

Over 20 m exposures of vertical sequences of stratigraphic succession are visible from the shore platform to the crestal part of coastal cliff section in Lakshmanpur-II village of the island. The windward side of southwest margin of the shoreline is noted for well preserved shore cliff sections and wide intertidal shore platforms

Fig. 2.68 **a** Wave-abrasive carbonate shore platform on the cemented surface (Neil Island) and **b** Mass-wasting process produced debris fall at the foot of the cliff (Neil west)

(Fig. 2.67). Stratigraphically, the Neil formation is well preserved in this part nearby the Natural Bridge. The Neil formation corresponds with Malacca Limestone and west coast limestone of Andaman and Nicobar group of islands and their age ranges from Pleistocene to Holocene. Singh et al. (1978) observed that the Neil formation can be subdivided into Round Formation, the Guitar Formation and the Malacca Limestone Formation. The coralline algae formation is associated with foraminifera and corals (Kishore et al., 2015). The two members of limestone formation are separated by unconformity in the stratigraphic succession. The upper limestone members constitute coralline algal assemblages with corals and shell limestone with coral rags. The lower members on the other hand constitute sandy limestones, silty limestones, mudstones or marls, and they belong to the ages of Miocene to Pleistocene period in such bedded formation of the carbonate island. The beds of these stratigraphic successions are dipping towards the sea at an angle ranging from 12° to 17° with presence of wide-spaced joints across the layers and number of seepage points along layers. Heaps of boulders size materials are deposited at the base of cliff wall due to rock falls or debris falls in the region. Number of fissure caves is also developed due to dissolution process in the massive limestones. Various joints and fault lines are identified across the western shore platforms of the region (Fig. 2.69).

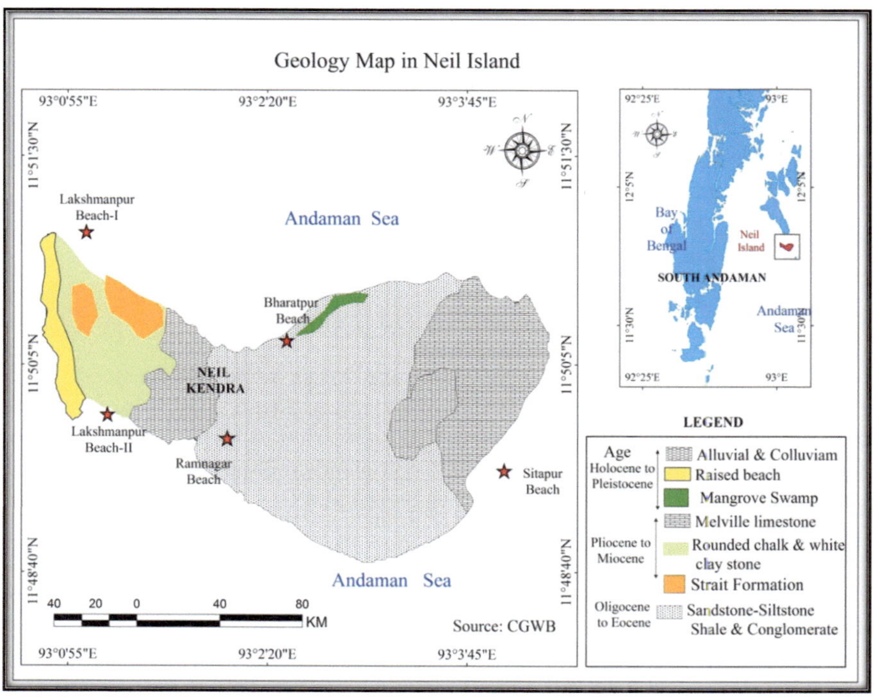

Fig. 2.69 Geological characters of Neil Island

Another attractive cliff section exposure with 25 mts elevation above sea level near Sitapur Beach represents stratigraphic succession of carbonate rocks. The section constitutes basement mudstones with overlying succession of chalky limestones, siltstones and clay stones with coral rags in the southeastern part of the island. Chalky limestones of the exposure site are affected by sub-vertical joints and development of extensive fissure caves. Rocks are also crushed and faulted in the western end of chalky limestone member of this stratigraphic succession. Large size boulders and debris materials are deposited at the base of the cliff wall by mass-wasting process (Fig. 2.68a) and subaerial weathering process in the region. The basement mudstone layer is exposed on the shore platform at the lower intertidal region and modern beach sands occupy the upper part of intertidal region in the form of unconsolidated sand sheets.

The leeward side (Fig. 2.68b) and embayed shores of Bharatpur, Lakshmanpur-I and portion of Sitapur section contains thick deposits of coralline sands, coral fragments, shingles, shoals, marine shells and other foraminifers. Siliciclastic sands are found in Sitapur Beach of Neil Island. Extensive beach rocks of calcareous sandstones and coralline sands with carbonate materials are found in the high-tidal and mid-tidal limits of Lakshmanpur-II and Ramnagar villages of the island. The relict features of reef terrace are visible along the southwestern shores of Sitapur in erosive remnant of fossil corals. However, the inner parts of Neil Island are occupied with coralline limestones, clay stones with coral rags and unconsolidated sands with coral rags. Mangrove clays with alluviums and carbonate muds are found in the form of wetland deposits in Ramnagar and Bharatpur areas of the island. The Central Ground Water Board (CGWB) has estimated the lithification rate of the rocks and carbonate deposits of Neil Island, Havelock Island and Little Andaman Island of South Andaman district.

2.6.3 *Little Andaman*

Geological structures and rock types determine the erosivity and weathering rate of the island platforms under suitable climatic zones. The oceanic island of Little Andaman system was evolved since the Miocene period to Holocene period in the Plate Tectonic setup of Island Arc origin. Geological structure of the oceanic island is well documented by Central Ground Water Board (2013, March) in its Scientific Report Series 'E' (Sharma and Kar, 2013).

Geologically, the island is separated into 2 major units and 4 sub units of formations. The overlying geological formations belongs to the Holocene to Pleistocene age and contain coralline sands and fragmented coral rags (minor), mangrove swamp clay, and another sub unit of raised beach materials on their margins. However, the underlying geological formations are grouped into Pleistocene to Miocene ages of materials and they contain Melville Limestone and Rounded Chalk and White clay

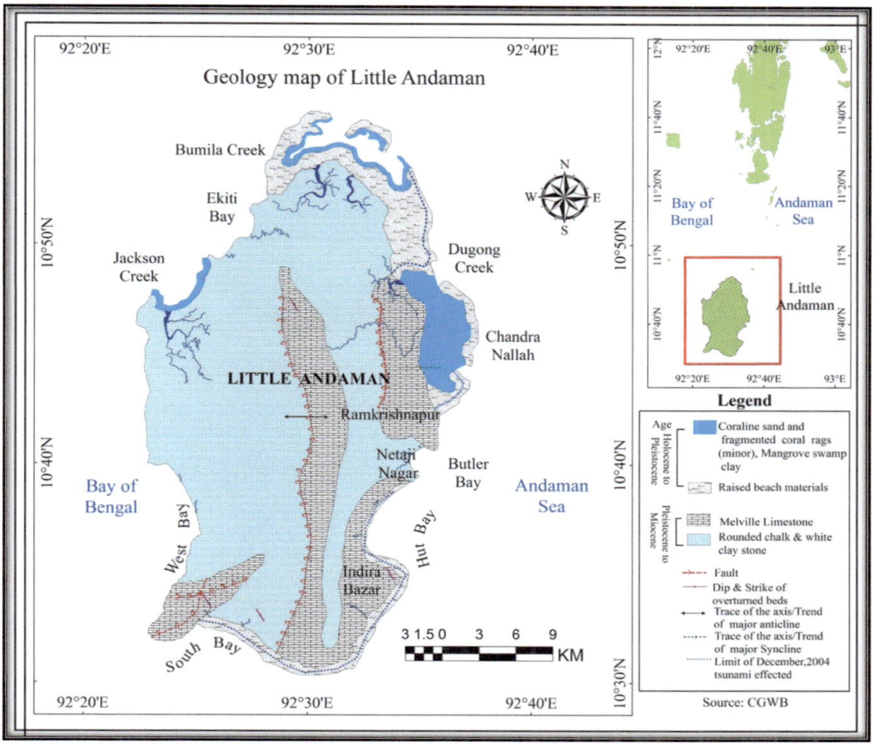

Fig. 2.70 Nature of geological diversity in Little Andaman

stone (Fig. 2.70). In another geological map of Little Andaman Island of the Geological Survey of India, 2012 (Bandyopadhyay and Carter, 2017) demonstrated about 4 different groups of geological formations of the island.

According to their demonstration of Geological Framework of the Andaman group of islands, the Little Andaman contains northeastern shore fringe Holocene sand, clay and coral reef formation; the northwestern, eastern, central and southern parts with chalk, mudstone, reworked tuff of Archipelago Group; southwestern shore fringe conglomerate and gritty sandstone of Mithakari Melange and a striking feature of southwestern shore fringe volcanics (Basalts) of deformed ultramafic. The island geological structures are also marked by number of N-S fault lines particularly on the eastern and southwestern shores and a series of synformal axis and anti-formal axis on the east.

Rocks of the geological formations are densely covered by forests and less affected by subaerial weathering and denudation processes. The exposed headlands with exposures of limestone rocks however, affected by severe weathering process and karstified (Fig. 2.71) significantly with developed geomorphological features of caves and cavernous deposits, tunnels, doline depressions, lappies, earth pillars and taffoni

Fig. 2.71 Pinnacle karst and karstified notched cliff on the black shale limestone at Kalapaththar, Little Andaman

marks. Limestone pavements on the shore platforms are wave-abraded and jointed by marine process.

2.7 Carbonate Settings of the Coasts

The submarine basin fringe deep water turbidites of sands and muds, ingraded conditions were uplifted, thrusted and deformed as sedimentary structures at the island fringes of the Andaman and Nicobar group in the active tectonic zones (Fig. 2.72). Such alternating layers of sandstones, shales, mudstones and grits of the island fringe environment are known as Andaman Flysch sediments. They are also mixed with carbonate sediments and reef limestones of coral rubble and debris materials at the shallow marine coral fringe settings. Today, at the shore cliffs, they are exposed in the forms of rhythmic, thin layers of sandstones, shales, mudstones and white claystone or chalks, particularly in the South Andaman group of Islands.

Gradually, with sea level oscillations and by active wave-cut processes, the Flysch sediments are wave-abraded and exposed on the shore platforms of the island margins. The shallow marine coral fringe reef sediments were deposited over the reef limestone pavements along the shore fringes during high tides. With time, they were lithified and cemented in a few areas of the shore platforms by the deposition of calcium carbonates, aragonites, dolomites and other micrites.

Beach calcarenites and aeolinites were also deposited around the high-tide zones and supratidal zones of the islands during the last glacial maximum (LGM), which were lithified and cemented as beach rocks and raised beach platforms over the Pleistocene reef terraces. Present-day beach sands, loosened reef debris, and other carbonate sediments occupied the shore platforms and upper tidal zones of the islands, but they were wave-reworked by the swell waves and storm surge waves (Fig. 2.72).

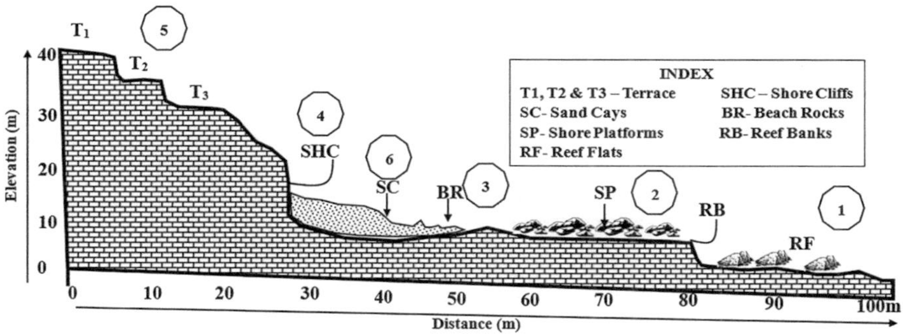

Fig. 2.72 Schematic diagram representing the ecological niches and topographic characters of the reef banks, reef flats, shore platforms, shore cliffs and island platforms of the South Andaman Island system

Thus, the entire shallow marine environment and the islands systems constitute, (i) fringing coral reefs with reef flats and banks, (ii) shore platforms of the limestone pavements with carbonate sediment deposits, (iii) upper tidal zones with beach rocks and beach calcarenite deposits, (iv) shore cliff turbidites with Andaman Flysch deposits, (v) reef terraces with eolitinites and calcarenites and (vi) sediment cays with beach ridges, bars and sand platforms as carbonate settings of the coasts.

Chapter 3
Karstified Coasts and Coastal Hydraulics

3.1 Karstification Process in the Islands

Karstification is the limestone dissolution process by rainwater, which played a major role in shaping the coral-fringed islands and their morphology in the Andaman and Nicobar Islands. The Andaman Islands receive 2982 mm of rainfall annually on their Andaman seaward sides (Port Blair), but the Little Andaman Islands receive 3500 mm of rainfall annually on the Bay of Bengal fringe areas. The island's systems are affected by rain during the northeast monsoon (December-February) and by rain during the southwest monsoon (May–October) in the Northern Indian Ocean. The limestone outcrops of the promontories, headlands, ridges, shore cliffs and shore platforms are karstified under subaerial weathering processes due to bicarbonate dissolution with the available amount of rainwater. Further, the island systems are covered by forest biomass, which adds the organic acids to mix with rainwater dissolved carbon dioxide, and when they mix with the $CaCO_3$ of the underlying rocks, that will produce calcium bicarbonate acids in dissolved conditions at the island fringes.

The dissolved bicarbonate acids will easily erode the limestone and other carbonate rocks at the island platforms, and sometimes they begin to dissolve the limestones of the shore platforms, promontories, headlands and the shore cliffs of the island fringe environment during rainy months, making them active to karstify the shores. The present-day shaping of the morphology of the coral fringe coasts by the karstification process played a major role in the evolution of the karstic features in the islands.

Rocks of shell limestones, reef limestones, white claystones, chalky limestones, marls, calcareous sandstones, beach and dune calcarenites with fissures, fractures, joints, as well as the algal encrustations, have played a significant role in the karstification process of the rock outcrops. Baratang limestone caves and caverns of South Andaman are developed on the thrust limestone ridges, which are intensely jointed and fractured and tilted, but rooted by rainforest vegetations and sharply affected by the rainwater dissolution process. The joint networks greatly facilitated

A. Paul et al., *Karstified Coasts of the South Andaman Islands*,
https://doi.org/10.1007/978-3-032-10299-7_3

the circulation of rainwater and drainage waters in the subterranean karst of the Andaman mainland ridges. The vadose caves of the region lie above the water table in the unsaturated vadose zone, but dissolution of limestones occurs during the rainy months and by the exposure of subterranean drainage networks. Cave formations and speleothems of Baratang limestone ridges are marked by dripstone deposits of stalactites, stalagmites and columns, extended tunnels, sinkholes and wider passages to the caves. Types of subterranean karst are also found in Little Andaman Island near 'Whisper' and 'White Surf' waterfalls at the cliff walls under the cover of dense rainforest vegetation (Fig. 3.1b, c).

Coastal karst landforms of the Ritchi's Archipelago, Little Andaman, Rutland Island, Neil Island and Havelock Island are remarkably diverse in the forms of phyto-karst or bio-karst, coral island karst, solution dolines, sea caves, flank margin caves, limestone solution notches, arches, limestone basins, corridor karst or giant grike-land, residual clint blocks, coastal karren, pinnacle karst and limestone pavements (Fig. 3.1a, c–f).

3.2 Classification of Karstified Coasts in Andaman Islands

The karstified coasts of South Andaman Islands are classified into two types: (i) Island Interior Karst and (ii) Coastal Karst is based on their location, elevation and origin. However, the karstic features of island interior karst are again classified into three sub-types (Fluviokarst, Limestone Dissolution Karst and Dissolution Non-dissolution mixing zone), similarly, the coastal karst is also divided into three sub-types (Pseudokarst, Phytokarst and Dissolution Non-dissolution mixing zone).

3.2.1 Karstified Coasts in Andaman

The carbonate coasts of Little Andaman, Havelock and Neil Islands evolved in a variety of types, shapes, origins and functions. A unique set of karst features has appeared in the soluble carbonates rocks along the seashores and island interiors. As soluble rocks are present immediately inland behind the sea face or along the banks of the river system, coastal processes have created distinctive caves and karst in these islands. Various types of caves and karst features are identified in the present work, but also observed that dynamic coastal processes constantly modify the sea caves, and obscuring the origin of caves in the region. Caves and karst features produced by dissolution processes, however, created a long-term preservational environment that helped us to explore the past (Fig. 3.2).

The carbonate coasts of South Andaman are located in the tropics and near equatorial environments, and consist of young and old carbonates. Both the hydrology and geochemistry of such coastal environments have created the caves and karsts of an identical form and character in the selected islands (Figs. 3.2 and 3.3). Broadly,

Fig. 3.1 Coastal karst landforms produced by the karstification process in South Andaman Island **a** residual clint blocks at Havelock Island, **b** giant grike land at Baratang Island, **c** drip stone runnels at the limestone cave surface of Baratang Islands, **d** limestone tunnels developed across the Kalapaththar limestone ridge at Little Andaman Island, **e** shore fringe pinnacle karsts of Kalapaththar, Little Andaman Island and **f** coastal karrens developed by karstification process but wave abraded in the shore platform of Neil Island

the caves are genetically classified as karst or pseudokarst caves. Karst caves are produced by dissolutional processes, and pseudokarst caves are created by non-dissolutional processes. Exposure of steep cliffs with a variety of lithologies represented Taffoni by sub-aerial weathering processes, and provided environmental conditions for the development of fissures with other types of pseudokarst caves and karst caves along the sea face of islands. Sea caves and sea arches are abundant in Neil Island, Little Andaman Island and Havelock Island, because geochemical

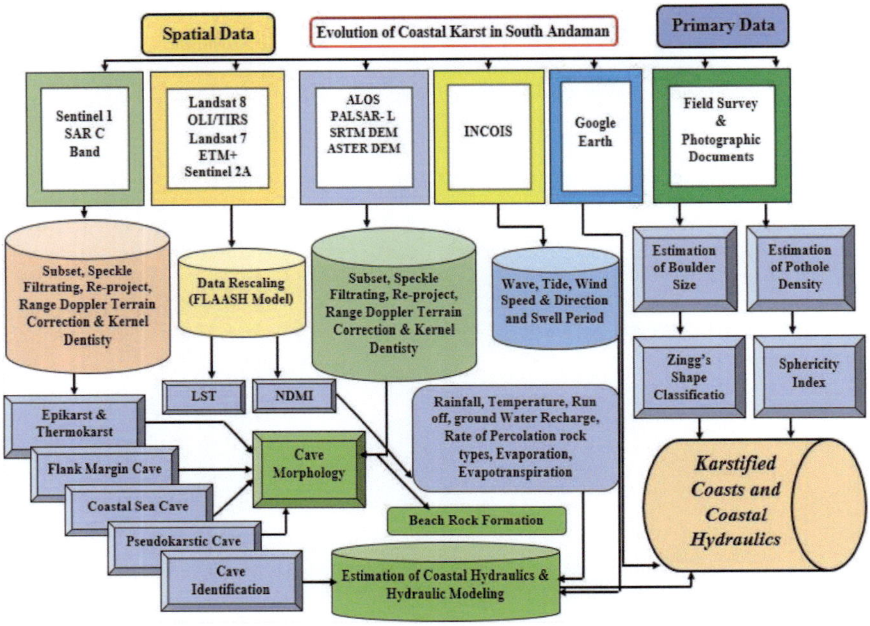

Fig. 3.2 Conceptual methodological flow diagram of the coastal hydraulic modelling

actions, weathering rates and wave abrasion processes are strongly acting over the carbonate rocks along the shore cliffs fronted with intertidal shore platforms.

A scheme of classification of the karstified coast (Table 3.1) is made for identification of various types, sub-types, features or morphology, and processes involved in the evolution of landforms using geospatial techniques with repeated field visits in the studied areas of South Andaman Islands.

Karstified coasts of all three islands are broadly categorized into Island interior karst and coastal karst for the scheme of classification. Again, the six sub-types of the two major types are considered based on process variation in the development of karst features along the shore fringes and distance inland from the shorelines. Both dissolutional and non-dissolutional processes are involved in the development of karst landscapes in the oceanic island coasts of the Andaman Sea and Bay of Bengal (Northern Indian Ocean). Favourable conditions are identified as island fringe coral reefs, carbonate rocks with jointed limestones, fractured rocks with fault lines, hydrology and drainage conditions, geochemical actions, high rainfall (> 3000 mm), dense forest cover areas with shore fringe phytography, strong evaporation rates, sub aerial weathering processes on exposure of the shore platforms, conduit flow to the sea ward sides, tidal pumping and active marine erosion-accretion processes. The presence of carbonic acids in the atmosphere, soil and water played a very significant role in the karstification of carbonated rocks and sedimentary landforms of the 'Archipelago Group' of Andaman Islands under a tropical climate (Fig. 3.3).

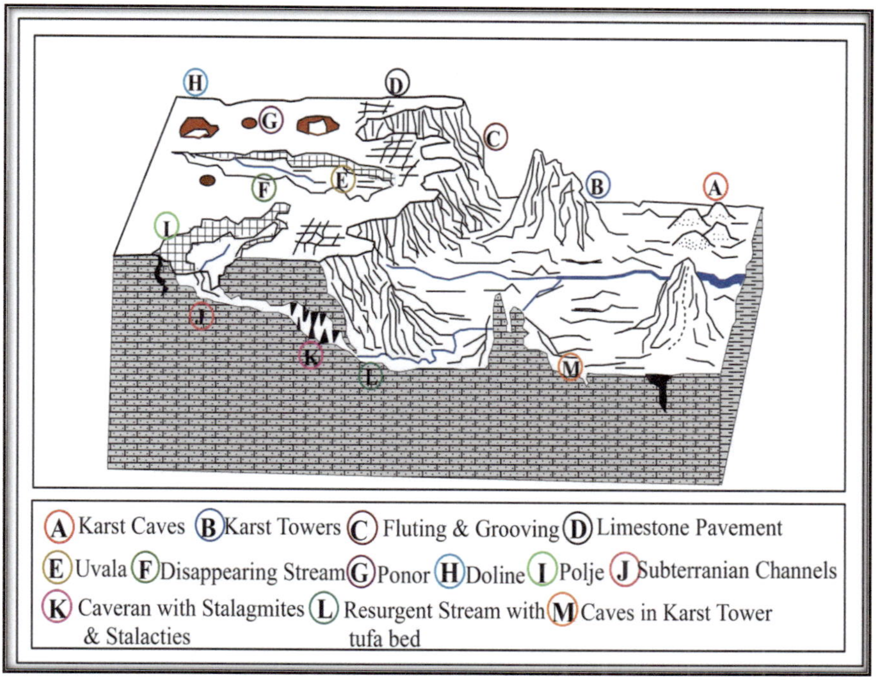

Fig. 3.3 The conceptual diagram of the surface karst formation

Extensive dissolution of aragonite, calcite, dolomite and halite generally occurs due to the action of carbonic acid derived from soil CO_2 of the island systems and atmospheric conditions of the region. The liquid bicarbonate acids derived from rain waters, rain forest organic matters and run off surface waters over the island platforms influence in rapid dissolution process of carbonate rocks in the islands.

Various karst features have been identified in different forms and origins (Fig. 3.3). These features are now explained in the following section with supporting evidences. Identification and delineation of karst landforms are generally based on satellite images (Sentinel 2 and 1, Landsat 5 ETM+, Landsat 7 and Landsat 8 OLI/TIRS), Google Earth, ALOS PALSAR DEM, SRTM DEM, Survey of India's Toposheets (R.F. 1:50,000), and repeated field survey works in three different islands.

3.2.2 Fluviokarst Features

Drainage channels across the terraced surfaces composed with carbonate rocks and lithified carbonate sediments have revealed various types of fluviokarst features in Little Andaman and Havelock Island. The soluble and fractured carbonate rocks of

Table 3.1 A classification scheme of karstic features for the South Andaman coasts

Schematic classification of the karstified coasts of Havelock Island, Neil Island and Little Andaman					
Sl. No.	Types of karstified coasts	Sub-types	Features/ Morphology	Processes involved	Comments
1	Island interior karst	I. Fluvio karst	Blind Valleys with resurgence and emergence; Abandoned limestone cliff with solutions pits and caves, springes with wall rock weathering pits; wrinkle marks on speleotherms; dripstone and flowstone speleotherms on the wall rock; valley heads in the karst rocks; extensive sub-terranean water drainage	Geochemical weathering of limestone; fluvial process	Drainage specific areas with channelled runoff space and surface lowering
		II. Limestone dissolution karst	Flank margin caves; karstic caves; sinkholes; caves speleotherms	Active dissolution process	Large scale surface lowering and recession of scarp
		III. Dissolution non-dissolution mixing zone	Doline depressions; deeply incised valleys; surface outcrops of karst rocks; grykes and clints	Partial solution; weathering of rock; surface lowering	Exposure of features on the surface; old stage of the karst cycle
2	Coastal karst	I. Pseudokarst (non-dissolutional)	Tafoni; limestone solution notch on a modern shore platform; limestone coastal karrien pitting on the coast; large scale clint-and-gryke terrain with corridor karst; pinnacle karst; limestone pavements on the shore platform; honeycomb weathering structure; lappies	Geochemical weathering on carbonate rocks; waves and tides; water layer weathering; salt weathering	Weathered and denuded surface

(continued)

Table 3.1 (continued)

Schematic classification of the karstified coasts of Havelock Island, Neil Island and Little Andaman					
Sl. No.	Types of karstified coasts	Sub-types	Features/ Morphology	Processes involved	Comments
		II. Phyto-karst	Karst basin on the shore platform with alge mat; coastal karrien pitting with alge mat; rock pools with sea grass patches on the shore platforms	Calcification with biochemical weathering process; evaporation on exposures; partial solution (wetting and drying)	Vertical surface etching with calcification and solution process
		III. Dissolution non-dissolution mixing zone	Unroofed caves; roofed caves with wide opening; fissure caves; talus caves; sea caves and sea arches; caves shaped by a combination of beach processes; solution runnels on the soft limestone chalk; solution flutings (rillen karren); notches	Non-dissolution process; partial solution process; speleogenesis; waves splashing	Formation of hybrid caves and cliff erosion

terraced surfaces result in the development of special karst landforms like depressions, sinkholes and courses of subterranean water drainage. The incised valleys of meander channels on the marshy terrain and highland surface have been delineated with the sections of blind valleys, dendritic valley heads with springs, points of resurgence and emergence and other karst features in the islands. Solution and stream action operate together in the drainage interfluves in the formation of such karst features. The leakages of drainage waters into the porous limestone terrain have resulted in the formation of dry valleys, dry storage basins and misfit valleys in Netaji Nagar and Ramkrishnapur villages of Little Andaman (Fig. 3.3). The karst springs are common throughout the region of south-central highlands, and they have been originated from perched stream caves at the flank margins of ridges and spurs (Fig. 3.3).

The 'White Surf' waterfalls (Fig. 3.4) and 'Whisper' waterfalls are within the distance between 4.5 km section of the drainage channel appears after crossing the abandoned limestone cliffs in the region under dense forest. Karst drainage systems of the islands under study are a key to understanding other karst features located in the surface ('epikarst') and near surface ('endokarst') zones (Fig. 3.5a). The storage and flows of such karst drainage systems are guided through the rate of precipitation, evaporation and discharge types. The majority of precipitation water discharges through surface overland flow, channel runoff, seepage flow, outflow springs, and evaporation rates and deep losses to regional ground water systems in the tropical islands. Karstification process is intensified in the tropical climate due to the nature and character of rock types and weathering rates. However, fluvial processes are significant in the formation of such surface and subterranean landforms. Fluviokarst features have also appeared in the northern and southern parts of Havelock Island over the marshy and forested tracts (Fig. 3.5b).

Fig. 3.4 Fluvio karst features developed on the bank margin of John Richardson Creek (Little Andaman)

Fig. 3.5 **a** Karstic cave on the limestone cliff and **b** A part of the dry valley section on the limestone terrain (Ramkrishnapur)

3.2.3 *Limestone Dissolution Karst*

Flank margin caves, karst caves, sinkholes and cave speleothems (Bögli, 2012) are important dissolutional karst landforms over the thick limestones and other carbonate rocks identified in the islands under study. Active dissolution process into the jointed limestones, geochemical weathering process into the fractured carbonate rocks, and presence of soluble minerals and carbonic acids have resulted in the development of such diverse landforms (Figs. 3.5, 3.6 and 3.7).

Many flank margin caves are developed in the freshwater lens on the scarpland margins or abandoned cliff walls immediately behind the coastal inland, and they are also called as karst caves, resulted from dissolution process under sea level high stands. The greatest amount of $CaCO_3$ dissolution occurs at the lens margin, and a large number of such caves are found on the simple carbonate islands (Havelock and Little Andaman) of the region. However, the position of lens margin moves with sea level changes, and accumulation of organics at these boundaries may also enhance dissolution that resulted in the formation of flank margin caves during the high sea level.

There are 15 flank margin caves in the Havelock Island, 5 in Neil Island (Fig. 3.6) and about 13 caves in Little Andaman Island, so far as the study is concerned. All the caves are now located at 30, 40, 70 and at the 100 m altitude above the mean sea level in the islands under study.

This indicates that the sea level was higher than the present, and during the Oligocene (33–23 M) to Pliocene epochs (5.3–2.58 M) it was over 23 m globally but the local upliftment was more significant to achieve the present position of flank margin caves in the island territory. The phreatic water chamber of the fresh water lens was higher with high sea level in the geological past when the flank margin caves originated in higher position of hill slope in the region with dissolutional process.

Fig. 3.6 A sea cave is developed along the cliff wall of the promontory of Neil Island

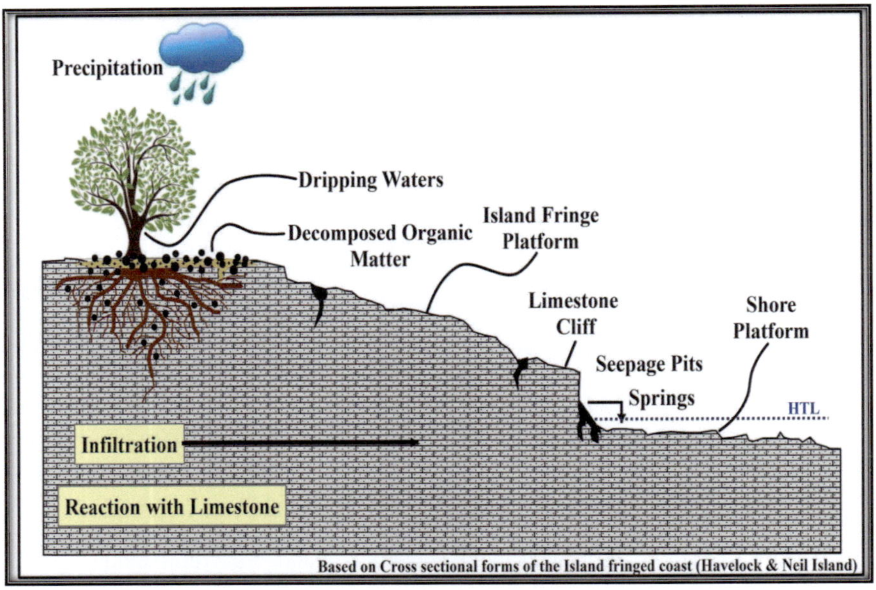

Fig. 3.7 Formative mechanism of flank margin caves

Flank margin caves of different altitude were formed and influenced by local tectonics and relative sea level changes with rise and fall of phreatic water zone and vadose water zone (Fig. 3.7).

3.2.4 Dissolution and Non-dissolution Mixing Zone

Small pits, solution pans, doline depressions and uvalas are found on the flat surfaces underlain by White Clay stones, Chalk and Melville limestones in the interiors of Little Andaman, Havelock Island and Neil Island by active dissolution process. The small pits are enlarged into solution pan, and medium size close depressions as dolines and large size elongated shaped close depressions as uvalas are developed in favourable conditions of karst environment of the region. Such features are easily delineated with the satellite images and field studies (Table 3.2).

The large-scale surface karst development is achieved in Havelock Island and Neil Island interiors over the chalk and white clay stones. They are mechanically weak and did not favour the formation of underground drainage but provided condition for evolution of medium scale to large scale surface karst landforms (Fig. 3.8). The frequency of surface karst formation is estimated as ground pits and sinkholes per square kilometre; and number of doline depressions and uvalas per square kilometre;

Table 3.2 Surface karst features measured in the Islands of South Andaman

Sl. No.	Surface karst features	Processes	Frequency per Sq. km		
			Havelock Island	Neil Island	Little Andaman
1	Pits and sinkholes	Solution	12	03	10
2	Ponor	Solution	06	02	08
3	Limestone pavement	Denudation and solution	07	02	10
4	Grikes/clints	Jointing and solution	18	–	16
5	Lappies	Solution	32	–	30
6	Disappearing stream	Seepage flow	0.5 km/km^2	–	0.8 km/km^2
7	Doline depressions	Solution and collapsing	1	1	2
8	Uvalas	Enlargement	1	1	1
9	Pinnacles and solution spikes	Solution	> 300	> 200	> 300
10	Solution pipes	Solution	5	6	7

Based on the field survey

ponor, limestone pavements, disappearing stream, grikes on limestone pavements and lappies in parts of the islands (Fig. 3.8).

Solution process on calcium carbonate dominated rocks is enhanced further by interaction of liquid carbonic acids generated with decomposition of organic matters in the dense forest cover areas and high rate of annual rainfall (over 3000 mm) in the islands of Archipelago Group (Table 3.2).

3.3 Coastal Karst in South Andaman Islands

Various coastal karst and mixing zone karst have been developed in Havelock, Neil and Little Andaman Islands.

3.3.1 Coastal Karsts

The coastal areas of Havelock Island, Neil Island and Little Andaman are fringed with various sub-types of karst features, i.e. the pseudokarst (non-dissolutional), phytokarst, and dissolution–non-dissolution mixing zone karst. In the coastal areas,

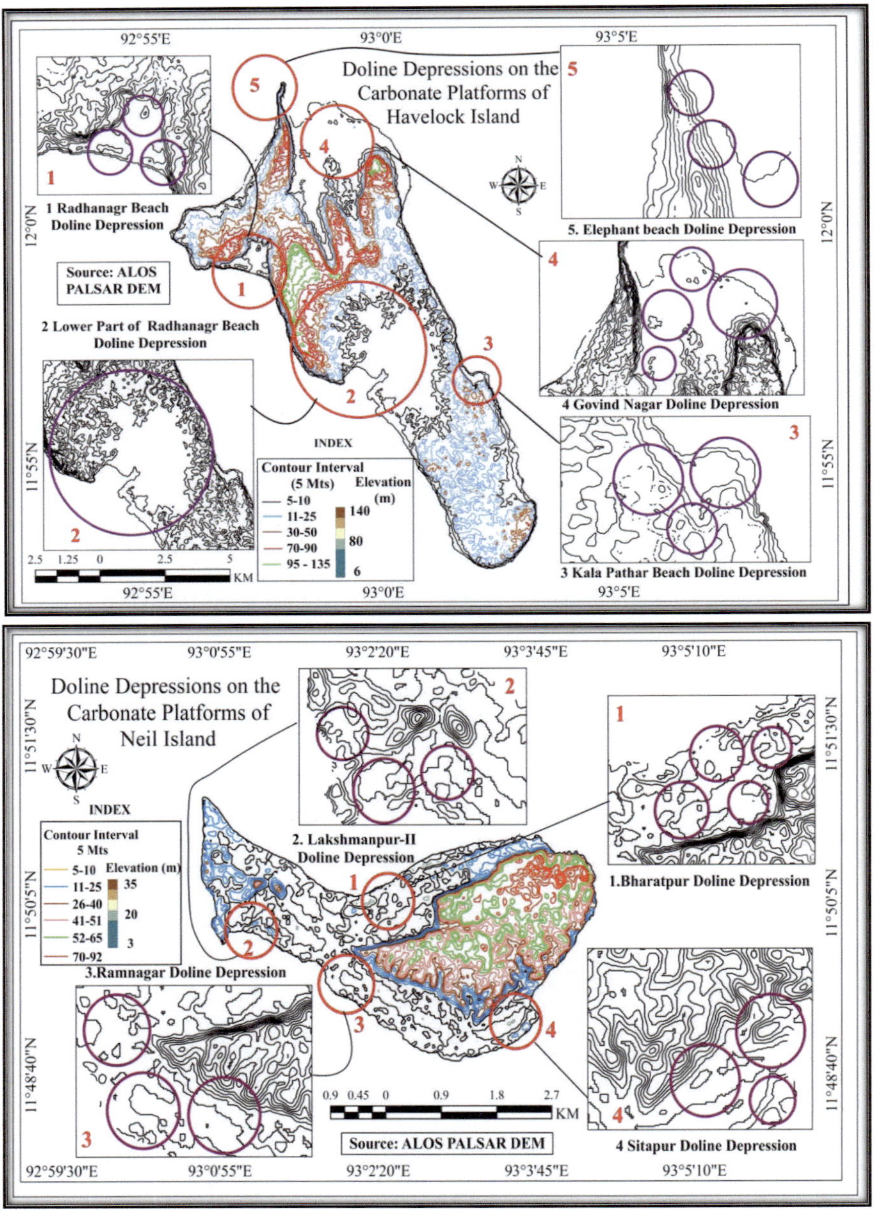

Fig. 3.8 Number of Doline depressions developed in the limestone surface of Havelock and Neil Island

Fig. 3.9 **a** Lappies developed on the limestone boulder in Ramnagar (Neil Island) and **b** Surface spikes and a groove channel in the shore platform of Ramnagar (Neil Island)

karst features are developed also by weathering, hydraulic action and tectonic disturbances. These features are not dominated by solution process in their fullest development and usually called as pseudokarst features. Shore platforms, shore cliffs, raised beach platforms and extended headlands towards the sea with underlain by soluble rocks provided ideal conditions for operating some processes in unique ways to produce characteristics features along the coastal areas. Several features of this group are identified as taffoni, limestone solution notch, coastal karren pitting, large scale clint and grikes terrain with corridor karst, pinnacle karst, limestone pavements, honeycomb weathering structure, lappies, etc. in the present study areas (Fig. 3.9a).

Portions of the rock that are relatively less affected by case hardening process are liable to chemical attack of salt weathering on the sea face. Salt accumulations, salt weathering and wave splashing on the cliff wall, cliff base rock debris, concave walls on the surface of overhanging hoods or visors, and also on the gentle sloping debris-strewn floors of the shore platforms have produced hollows or cavities on the rock surface. Such features are known as taffoni and they are concentrated along the upper parts of shore platforms due to operation of wetting and drying process under frequent inundations and exposures of tides. It is indicated by field evidences that taffoni (Fig. 3.9b) affected rocks are protected from rain wash process during monsoon months and they have become the areas of salt accumulations and further concentration of salt weathering activities.

The basal parts of limestone cliffs in many places are affected by wave hydraulics action, encroachment of tidal limit on the landward margins, and by solution process during exposures. They are known as wave-cut cliff base notches in limestones and calcareous sandstones of the island coast. The amount of graveliferous sediments pushed on the cliff base takes part in the wave abrasion process for further extension of notches during high tides in the region. Field evidences suggest the cliff base notches of Lakshmanpur and Sitapur shoreline of Neil Island, Kalapathar and Radhanagar Coast of Havelock Island, as well as other areas like Butler Bay, Hut Bay and Harminder Bay shoreline of Little Andaman showed rapid rate of erosion during tsunami waves, swell waves and storm waves (Fig. 3.10). Intertidal notches and sub tidal notches are very common in the shore platforms of Neil Island and Little Andaman Island.

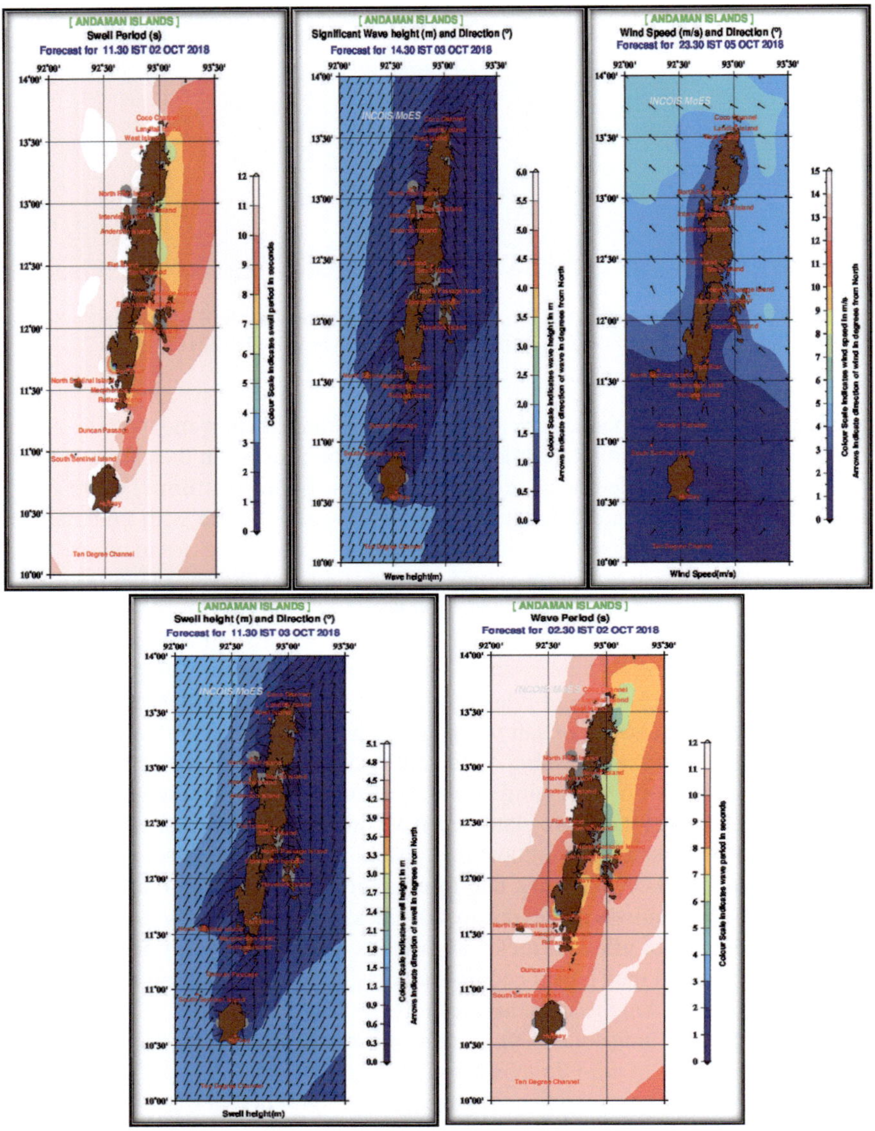

Fig. 3.10 The real-time data of the sea waves and swell waves on the sea face of South Andaman Islands

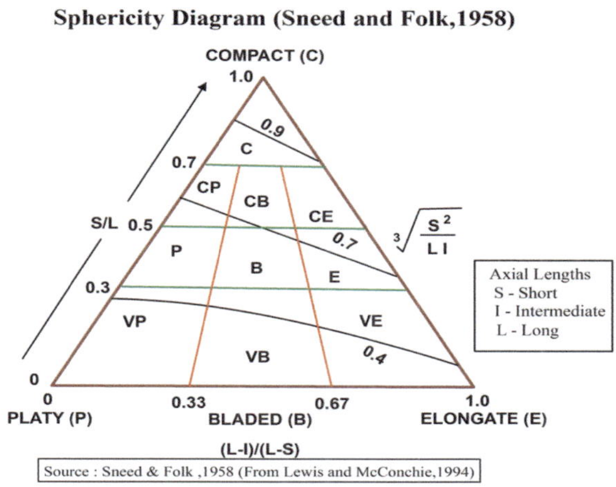

Fig. 3.11 Representing the spherecity index of shore platform-based pot-holes by Sneed and Folk, 1958

Appearance of pits, pans, micro pits and spikes on the upper part of shore platforms displays distinctive solution topography along the coastlines of Havelock, Neil and Little Andaman (Fig. 3.11). The carbonate shore platforms with calcium carbonate sedimentary surface and overlying algal mats helped to form coastal karren features by weathering, wave hydraulic action, wetting and drying, hydration and salt weathering process. Boring and grazing organisms over the carbonated rocks provided conditions for the formation of pitting, micro pits and pans (Table 3.3).

Gradually, the surface pittings have been converted into developed spikes with pan-like depressions by salt weathering and solution processes on the shore platforms (Ellison, 1994; Ford & Williams, 2017). Larger size potholes, like depressions, are found to develop on the limestone surface by wave action with the friction generated by loosened rock debris in the stage of rising tides along the shorelines. They become circular to semi-circular in shape or may be modified into saucer-shaped depressions by further weathering and hydraulic actions over the calcareous sandstones and limestones. They are flat-floored and steep sided and ranging from 10 to 48 cm in depth and 16–39 cm in length and 13–30 cm in width around the shore platforms of Neil Island and Havelock Island. Sometimes these features of surface karst are also known as phytokarst in the tropical coast. The spherecity index values of pot holes are calculated on the basis Sneed and Folk's Model, 1958 (Fig. 3.11).

The pools of water bodies with moist surface and sea grass patches have been expanded with chemical weathering, wave actions and coalescing process to form the rock basins into compound forms in the middle part and in the lower parts of shore platforms. Such rock basins hold pool of water bodies and provided conditions for sustained patch reefs and sea grasses in the low tide platforms of Neil Island and Little Andaman Island (Fig. 3.14). The bedding planes, joints and fault planes on the limestone cliffs are all vulnerable to chemical attack and erosion attack with various agents. However, where waves attack a promontory base from both sides, a hollow may develop on the areas of geological weakness along the carbonate coast to form

Table 3.3 The database tabulated on the calculation of spherecity index of pot-holes on the basis of field survey records (2012–2018)

Points	Latitude	Longitude	Length(cm)-L	Width (cm)-S	Depth (cm)-I	(L-I)/(L-S)	S/L	$\sqrt[3]{s^2/LI}$
Havelock Island (Kalapathar Beach)-Pothole-I								
PH-1	11° 58′ 34.4″ N	93° 01′ 08.3″ E	105	76	35	2.4137931	0.7238095	1.1626704
PH-2			60	53	28	4.5714286	0.8833333	1.1869001
PH-3			29	48	23	− 0.3157895	1.6551724	1.5116534
PH-4			39	30	10	3.2222222	0.7692308	1.3214761
PH-5			76	60	31	2.8125	0.7894737	1.1517965
PH-6			140	62	35	1.3461538	0.4428571	0.9222792
PH-7			90	31	33	0.9661017	0.3444444	0.6865239
PH-8			62	60	45	8.5	0.9677419	1.088678
PH-9			105	73	48	1.78125	0.6952381	1.0187596
PH-10			37	26	18	1.7272727	0.7027027	1.0049802
Neil Island (Ramnagar Beach)- Pothole-I								
PH-1	11° 49′ 21.2″ N	93° 01′ 54.4″ E	36	30	23	2.1666667	0.8333333	1.028184
PH-2			38	32	28	1.6666667	0.8421053	0.987308
PH-3			37	25	21	1.3333333	0.6756757	0.930007
PH-4			30	22	16	1.75	0.7333333	1.00277
PH-5			30	20	18	1.2	0.6666667	0.904806
PH-6			30	19	9	1.9090909	0.6333333	1.101661
PH-7			90	10	60	0.375	0.1111111	0.264567
PH-8			60	3	45	0.2631579	0.05	0.14938
PH-9			50	25	9.9	1.604	0.5	1.080832
PH-10			35.9	13	7.7	1.231441	0.362117	0.848725

(continued)

Table 3.3 (continued)

Points	Latitude	Longitude	Length(cm)-L	Width (cm)-S	Depth (cm)-I	(L-I)/(L-S)	S/L	$\sqrt[3]{s^2}/LI$
Havelock Island (Kalapathar Beach)-Pothole-I								
Little Andaman (Kalapathar Beach)- Pothole-I								
PH-1	10° 39′ 30.23″ N	92° 34′ 55.41″ E	30	27	16	4.6666667	0.9	1.1494641
PH-2			65	61.5	32	9.4285714	0.9461538	1.2205689
PH-3			25	27	40	7.5	1.08	0.9
PH-4			60	53	28	4.5714286	0.8833333	1.1869001
PH-5			12	8	20	-2	0.6666667	0.6436596
PH-6			90	31	33	0.9661017	0.3444444	0.6865239
PH-7			62	60	45	8.5	0.9677419	1.088678
PH-8			30	22	20	1.25	0.7333333	0.9308893
PH-9			40	28	20	1.6666667	0.7	0.9932884
PH-10			67	63	54	3.25	0.9402985	1.0313455

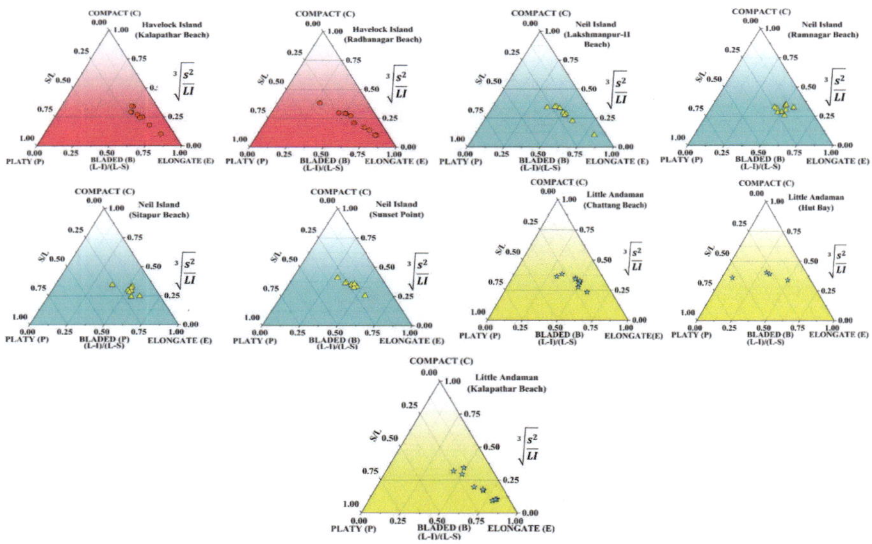

Fig. 3.12 Spherecity index values plotted and categorized for pot holes, Little Andaman

a sea arch. Both the arch roofs and sea tunnels are well developed by wave actions and chemical weathering processes along the coastal cliffs of Neil Island and Little Andaman Island. As long as the wall rock has sufficient strength to hold the roof of sea tunnel perpendicular to the slope, the natural arch survives in the carbonate coast. There are two different natural arches in the promontories of Neil Island at present. However, a number of sea stacks are formed in the island shores after the toppling of natural arches (Fig. 3.12).

Vertical to sub-vertical faults on the sea cliffs of white clay stones, chalk and dolomitic limestones of all three islands have been expanded by chemical weathering attacks. The narrow and steep-sided fault planes or joints after weathering and denudation process on the cliff wall produce spectacular cleft (Fig. 3.13b). Such fault clefts are found on the sea cliffs of Neil Island (Lakshmanpur and Sitapur) and Havelock Island (Radhanagar East, and Elephant beach point) in the region of Archipelago formation. Structural weakness of the wall rocks, rain water seepage through the joints or fault planes, and run off water concentrations along the expanded joints in rocks from the island slopes provided condition for such development of frequent clefts (Fig. 3.13a).

In parts of the intertidal shore platforms of Little Andaman, Havelock Island and Neil Island grikes have grown on the limestone pavement to form a typical topography of crisscrossing corridors (Fig. 3.13b). Such large-scale grikes are called also as corridor karst in the tropical coast. Recession of such grike-walls with wave erosion and salt weathering process have produced several residual clint blocks in the flat lying limestone platform near Sitapur, Kalapathar, Radhanagar, etc., sections of the island shorelines in South Andaman.

Fig. 3.13 **a** The natural arch of Neil Island **b** Development of the cleft on the foot wall of the shore cliff

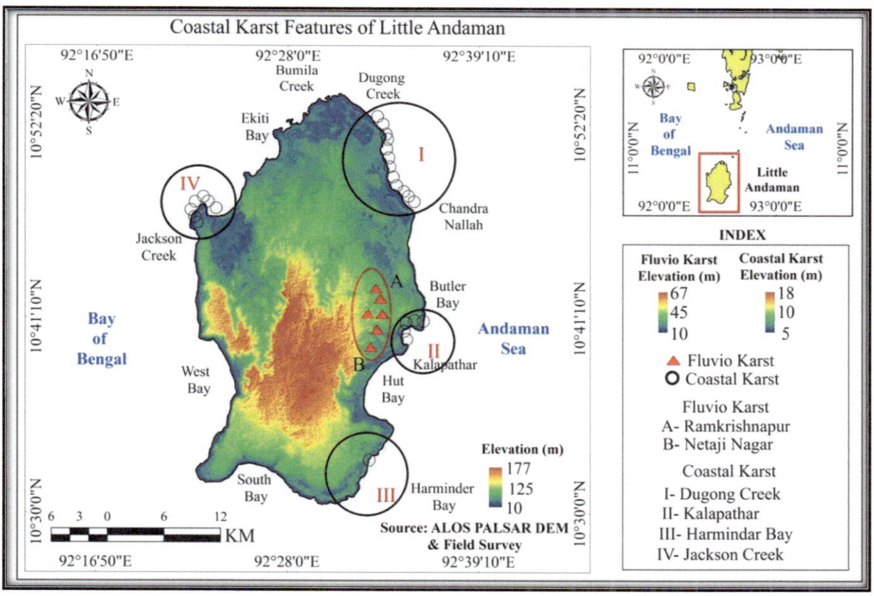

Fig. 3.14 Diversity in coastal karst features in Little Andaman

Spectacular development of pinnacle karst is found in the shore fringe areas of Little Andaman. The needle like pillars stands 1.6–30 m tall with diameters of 1–15 m along the residual limestone cliff in the Hut Bay, Buttler Bay and Harmindar Bay of the island. Development of such feature was possible due to the general erosion of regional cover sediment from the top of the shore cliffs in the areas of the coast. Gradually, by limestone solution they seem to have formed and enlarged under forest covered corridor with concentration of carbonic acids in the tropical island shores (Fig. 3.14).

The weathered limestone surface usually found in the region of coastal karst along the island shores consisting of fluted, etched and pitted rock pinnacles in the smaller forms are separated by deep grooves (Fig. 3.13). The water containing carbonic acids and humic acids penetrate along the joints of limestone rocks of the carbonate coast, and expand in the form of etched or rugged surface by solution of rock on the shore platforms. Such fluted forms of bare rocks are known as 'lappies' and commonly found on the coralline limestone blocks and also on the moisture surface of shore platform rocks with thick mat of algae. In dry condition, the limestone base and shore platform rocks become extremely hard in the island shores. Diverse karren features including lapis and shore platform karst features of the island coasts also may be termed as phytokarst landforms in different scales of development.

3.3.2 Mixing Zone Karst

The simple carbonate island of South Andaman exhibit extensive dry caves at the extreme high tide level (HTL) of Neil Island, Havelock Island and in the parts of Little Andaman. Coastal cliffs and promontories are the identical sites in which such caves are found. They are formed during the past glacio-eustatic high stands of sea level but presently lying at the base of lower sea level than earlier position. The fresh water lens within the carbonate rocks provide two separate mixing zones for dissolution and mixing dissolution environment. When the vadose water descends through the underlying carbonate rocks and mixes with the phreatic water lens they contain saturated water with $CaCO_3$ to produce active dissolution process on the top of fresh water lens. Similar conditions also develop in the bottom of the fresh water lens in which fresh water mixing with marine water, and appearing vulnerable to CO_2 degassing in the region but the condition counter-acts the environment of mixing-dissolution process. As the islands are covered by dense vegetation in the tropical environment the decomposed organic matters leach through the jointed rocks to reach the fresh water lens margins, and they provide the geochemical activity of included organics at their boundaries. Under such condition, the greatest amount of $CaCO_3$ dissolution occurs at the fresh water lens margins of the simple carbonate islands. Larger caves at the base of the cliffs are known as 'flank margin caves' because of their location along the margin of the lens, and probably formed by dissolution in the earlier period but modified by wave actions and salt weathering process at present (Fig. 3.14). Years of field work (2012–2020) in the islands of South Andaman and in other localities has revealed that the larger caves are found on small and simple carbonate island of the region.

The flank margin (Mylroie et al., 2017) caves initially develop as a small phreatic chamber to large phreatic chamber at the fresh water lens and finally convert into the elaborate phreatic caves with dissolution and by wetting and drying process under tropical hot and humid climate. The opening passage of flank margin cave is 17–20 m wide and 10–15 m in height along the shore cliffs of Neil Island. However, in Radhanagar east (Havelock Island) the caves are relatively smaller and ranging from

Fig. 3.15 Wave cut notches on the foot slope of the limestone cliff fringed with the shore platform

10 to 12 m in width and 3–5 m in passage height. The higher passage in Butler Bay caves (Little Andaman) are also above 10 m in elevation. Larger flank margin caves are also seed on the sea face of south Bay shoreline of Little Andaman (Fig. 3.15).

There are other types of Pseudokarst caves, i.e. fissure caves and talus caves, and the most common cave found in the coastal environment is the sea cave particularly in and around the present islands of South Andaman. On the steepened hill sides and cliffs, and also on promontories the internal earth forces or tectonics have created fractures and sub-vertical joints. Such fissures on the cliff wall are widened and deepened by solution, wave actions and salt weathering processes along the shorelines of Havelock Island and Neil Island. Fissure caves are constantly being destroyed and replaced by the new fissures with the shore cliff retreated due to coastal erosion.

The collapse materials from cliff tops produce voids by limestone blocks on the shore platforms and create linear cave systems. As the large blocks material in the form of talus can persist sufficiently long enough on the coastal environment to create cave systems, they are termed as talus caves. In few cases wave actions remove or reduce the basal debris collapsed from the cliff failure and convert them into short-lived cave system along the upper part of shore platforms. Ample of such caves are also visible along the cliff base of Neil, Havelock and Little Andaman Islands.

Varieties of sea arches and sea caves in smaller and larger forms are usually found along the littoral tracts also known as 'Littoral caves' in the carbonate coast. Sea level changes, soluble rocks, weathering activities and wave actions promote condition for development of such cave systems. The present study areas demonstrate both the arches and sea caves along the cliff walls and also along the bank margins of seaward edge shore platforms in Andaman by the events of larger marine inundations with significant wave heights. Field visits reveal that sea caves vary from arches or natural bridges and small voids to large chambers up to 15 m wide, 12 m deep and 9 m elevation in the tropical oceanic islands.

The windward side of steepened bank margins on the seaward edge of the shore platforms represent unique formation of buttress zone with reef front in places of Neil Island and Little Andaman. The reef front buttress zones on the windward side provide most ideal areas of wave energy concentration. The carbonate rocks of the

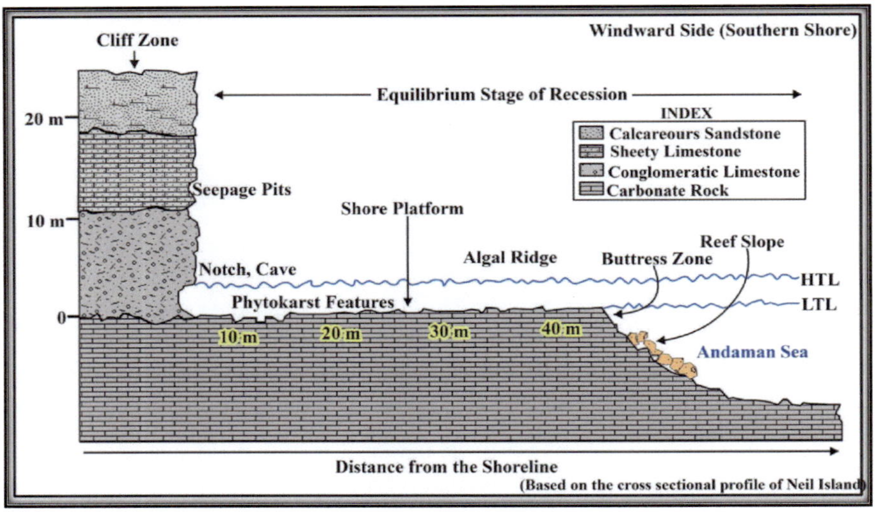

Fig. 3.16 Constant width of the shore platform due to similar rate of cliff and buttress wall recessions in the carbonate coast of Neil Island

reef front buttress zones are deeply scoured by wave actions and modified into spurs, grooves, tunnels, natural bridges or arched in the smaller scale with water layer weathering and wave abrasion processes. These features are clearly visible on the sea ward edge of the shore platform in the low tides in Lakshmanpur-2 section of Neil Island (Fig. 3.16).

3.4 Cave Morphology in the Islands

The morphology of caves is described in the next section to explain the morphological characters of diverse cave types in the tropical coast of South Andaman Islands.

3.4.1 Karstic and Pseudokarstic Caves

The present study areas have displayed caves produced by karst (dissolutional) processes, and other types of caves produced by pseudokarst or non-dissolutional processes (Fig. 3.17). The Andaman Islands limestone are weathered by dissolution in rain water which contains dissolved carbon-dioxide to form carbonic acid (H_2CO_3) to continue the reaction in the forms of above equation.

$$CaCO_3(s) + H_2CO_3(aq) = Ca_2 + (aq) + 2HCO_3(aq)$$

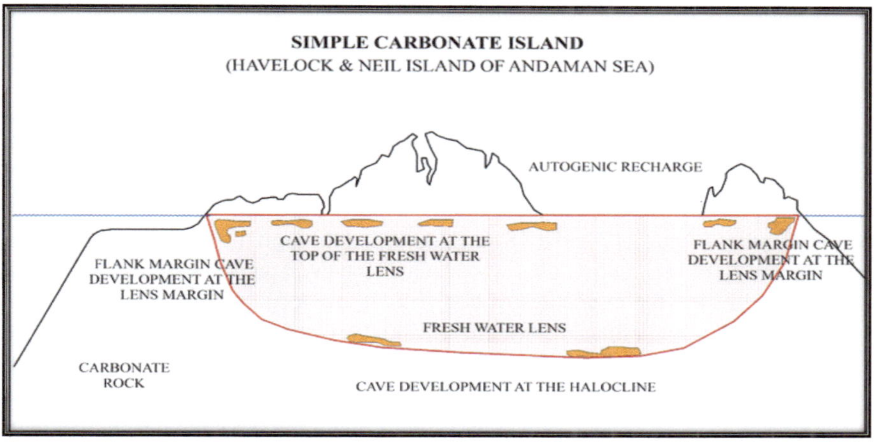

Fig. 3.17 Types of caves produced under different mechanisms

Thus, the reaction can explain the mechanisms of formation of underground caves. As the islands are densely covered by evergreen forests in the tropical hot and humid climate, the higher temperature gives rise to greater biological activity to produce the large amount of carbon-dioxide. The existing limestones are soluble to a limited extent in water but, dissolved carbonic acids increases the rapid rate of weathering and erosional activities of the soluble rocks in the inner parts of the island dominated by hill ridges that produce caves at the hill bases.

3.4.2 Epikarst and Endokarstic Features

Considering with karst drainage system of storages and flows entire karstic features can be grouped under Epikarst and Endokarst (Palmer, 1991, 2007). The Epikarst features are developed over the weathered surface of topsoil after surface detention of rain waters as well as their point recharge and diffuse recharge across the jointed rocks. The south central basin of Havelock Island between eastern and western coastal hills, and the valley flats of northern embayed shores of the same island exhibits Epikarst features (e.g. Pinnacle karst, Clints and grikes on limestone pavements) under the marshy tracts. The calcareous mudstones, clay stones, sandstones, and limestones were existed as karst bedrock beneath the soil. Jointed rocks of above category were largely weathered by dissolution process through the percolation of rainwater into the basinal surface of the island. Rainwater becomes acidified with the mixing of atmospheric carbon dioxides, calcified soils and decomposed organic matters of the forested landscapes, and thereby producing liquid bicarbonates and carbonic acids, on percolating downwards into the bedrocks from surface detention, they acted as powerful agent of dissolution process in the region. After heavy rain in the monsoon months (May–October) the saturated zone of Epikarst is well developed in the basinal surface, and thereby causing weathered surface of topsoil to develop the pinnacle karst in the region (Table 3.4).

The clayey soils between the jointed bedrocks and in between the pinnacle spikes with mats of grasses and spongy micro algaes hold moistures even in the dry months (November–April) to enhance weathering process. Such pinnacle karst features are well developed in Havelock Island under the marshy tracts which are documented in the field studies (Figs. 3.17 and 3.18) and identified through the Google earth images and vigorous field verification over wide areas. However, on the outer margins of the south-central basin, the exposures of bedrocks in the form of limestone pavements with joints, grikes and clints are evidenced, and thereby indicating as residues of the bedrocks. The presence of solution pits and solution hollows are also exhibiting the spread of karst terrain on the basin margin surfaces of the region as Epikarst features.

Endokarst features are actually consisting of major part of karst terrain extended into the underground areas of carbonate rocks and into the entire cave systems of the islands of tropical climate. Such features are situated below the Epikarst zone and in which the karstification is sufficiently matured (Ford & Williams, 1989). The islands of limestone terrains and other carbonate rocks are densely forested with evergreen vegetations and the system is operated by a large amount of rainfall (> 3500 mm) and high temperatures (> 30° C) in the tropics.

As the acid water from the vegetated soils and rocky surfaces can reach the deeper parts of karstic rocks through active dissolution process, the Endokarst features can develop in the form of dolines (Fig. 3.20); sink holes, caves in the undergrounds. Both the humic acids and carbonic acids generated through the biological activities from the forested landscapes and through the exchanges between atmosphere and rainwater took part a significant role in dissolution of carbonate rocks in the oceanic islands (Figs. 3.19, 3.20 and 3.21).

The calcareous rocks and limestones of the island are highly jointed, fractured fissured and faulted in many places that allowed the flow of acid waters and suspended materials into the deeper parts of stratification. Various types of karst holes, dolines, sea caves, fissure caves and caves in the karst hills are developed as Endokarst features in the carbonate islands (Tables 3.4, 3.5 and 3.6). Endokarst is governed by the computation between processes of dissolution, carbonate precipitation and clogging with non-dissolved particulate matters below the saturated zone (Rodet et al., 1995) of Karstified rocks (Figs. 3.18, 3.19 and 3.20).

3.4.3 Phytokarst, Zookarst and Other Processes

3.4.3.1 Phytokarstic Process

Karstic landforms of the islands generated and influenced by organic processes or biological processes can be grouped under biokarst features. These features may be developed due to the activities of plants or micro-organisms and also by the action

Table 3.4 Various karstic and pseudokarstic caves of the South Andaman Islands identified in the field survey (Havelock Island)

Havelock Island					
Location	No of caves	Latitude	Longitude	Elevation (m)	Rock types
A.*Hybrid Caves (Intertidal Coastal Zone)*					
Radhanagar Beach	8	11° 57′ 10.55″ N	92° 58′ 18.46″ E	2.5	White Clay stone
		11° 57′ 29.47″ N	92° 58′ 8.33″ E	2.5	White Clay stone
		11° 57′ 45.38″ N	92° 58′ 2.94″ E	2.5	White Clay stone
		11° 58′ 0.95″ N	92° 57′ 57.89″ E	2.5	White Clay stone
		11° 58′ 11.92″ N	92° 57′ 58.70″ E	2.5	White Clay stone
		11° 58′ 19.75″ N	92° 57′ 48.90″ E	2.5	White Clay stone
		11° 58′ 35.86″ N	92° 57′ 51.31″ E	2.5	White Clay stone
		11° 58′ 50.37″ N	92° 57′ 37.35″ E	2.5	White Clay stone
Elephant Beach	4	12° 2′ 7.23″ N	92° 57′ 47.88″ E	5	Mud Stone
		12° 2′ 14.00″ N	92° 57′ 50.01″ E	5	Mud Stone
		12° 2′ 36.26″ N	92° 57′ 47.45″ E	5	White Clay stone
		12° 2′ 47.04″ N	92° 57′ 51.80″ E	5	White Clay stone
Kalapathar Beach	2	11° 58′ 54.50″ N	93° 0′ 56.79″ E	2	Shaley Limestone
		11° 58′ 47.25″ N	93° 0′ 58.28″ E	2	Shaley Limestone
Southern part of Havelock	6	11° 53′ 59.40″ N	93° 3′ 11.46″ E	3	White Clay stone
		11° 53′ 31.39″ N	93° 3′ 3.13″ E	3	White Clay stone
		11° 53′ 13.19″ N	93° 2′ 50.07″ E	3	White Clay stone
		11° 53′ 2.66″ N	93° 2′37.68″ E	3	White Clay stone
		11° 52′ 56.69″ N	93° 2′ 15.72″ E	3	White Clay stone
		11° 53′ 4.08″ N	93° 1′ 52.87″ E	3	White Clay stone

(continued)

Table 3.4 (continued)

Havelock Island					
Location	No of caves	Latitude	Longitude	Elevation (m)	Rock types
B.*Flank Margin Caves (Inland from the Coast)*					
Radhanagar Beach	1	11° 59′ 10.46″ N	92° 56′ 42.30″ E	35	White Clay stone
Elephant Beach	3	12° 0′ 6.05″ N	92° 56′ 16.30″ E	30	Mud Stone
		12° 2′ 15.26″ N	92° 57′ 55.71″ E	20	Sandy Limestone
		12° 2′ 3.28″ N	92° 57′ 58.34″ E	20	Sandy Limestone
Beside Havelock Ferry Ghat	5	12° 1′ 45.78″ N	92° 58′ 0.11″ E	20	Sandy Limestone
		12° 1′ 34.57″ N	92° 58′ 1.93″ E	20	Sandy Limestone
		12° 1′ 22.29″ N	92° 58′ 3.78″ E	20	Sandy Limestone
		11° 59′ 47.33″ N	92° 58′ 33.51″ E	72	Clay Stone
		11° 59′ 33.03″ N	92° 58′ 38.75″ E	72	Clay Stone
Middle Part of Havelock	4	11° 59′ 18.51″ N	92° 58′ 46.17″ E	72	Clay Stone
		11° 59′ 21.48″ N	92° 59′ 2.46″ E	72	Clay Stone
		11° 59′ 41.00″ N	92° 58′ 59.29″ E	72	Clay Stone
		12° 1′ 5.66″ N	92° 59′ 40.53″ E	20	Sandy Limestone
Near Bijay Nagar Beach	2	12° 0′ 53.18″ N	93° 0′ 9.62″ E	20	Sandy Limestone
		11° 56′ 49.93″ N	93° 2′ 8.78″ E	20	Shaley Limestone

of rock boring animals. Phytokarstic processes are largely influenced by the action of cyanobacteria on the karstic rocks. The blue-green algae in attached forms on the limestone rocks and other calcareous rocks of the shore platforms, swampy terrain of the island interior valley flats and basins played the range of active and passive roles in geochemical transform actions, and also in re-precipitation of calcites (Figs. 3.22, 3.23, 3.25 and 3.26).

Limestone pinnacles and coastal karrens are well developed in the moisture dominated shore platforms under the thick blanket of blue-green algae, in which the weathered surface is dissected in a random sponge work pattern with sharp edged spikes and has been blackened over the surface after evaporation. Such formation of fretted surface by endolithic perfarant algae over the limestone and other carbonate rocks are termed as Phytokarst (Viles, 1988).

Fig. 3.18 Tunnel, taffoni mark, stone pinnacles and stack

It has generated a ferociously sharp, intricate and spongy rock surface along the shore platforms of Neil Island, Havelock Island, and parts of Little Andaman. The exposure surfaces of Mileville limestone of Little Andaman under the swampy terrain adjacent of Dugong Creek exhibit such development of pinnacled karst (Figs. 3.23, 3.26, 3.27).

3.4.3.2 Other Bio-karst Features (Zookarst)

Other than phytokarst, boring fungi and lichens, rock boring organisms, boring algae, boring bivalves and boring barnacles also played significant role in bio-erosion, bio-chemical weathering and boring features development over the limestone rocks and other calcareous rocks of the sea shores. The carbonate rocks of the shore cliffs and large scale drifted boulders of the shore platforms are attached with the fungal portions of lichens as they have penetrated into the rocks. They usually exploit the weakness of calcite crystal interface and make the signatures of large spits on the rock surface (Fig. 3.24a).

There is a strong relationship between morphological zonation and biological zonation across the intertidal shore platforms. Trudgil (1987) proposed that there is a cause-and-effect relationship between biological zonation and morphological zonation of intertidal shore platforms. The drifted log-woods and the softer rocks (white

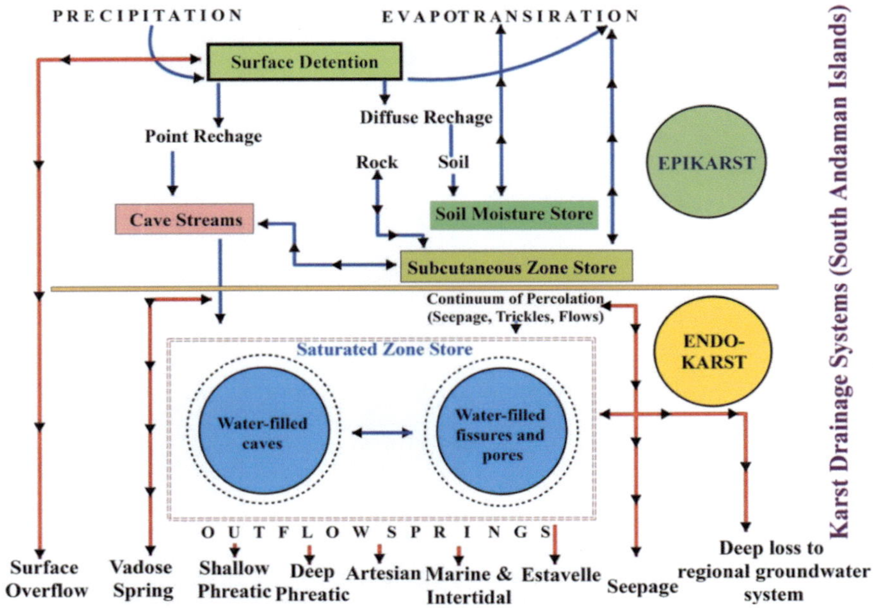

Fig. 3.19 A schematic representation of the different forms of Epikarst and Endokarst features produced by water flow

clay stone, marl and chalk) are targeted by the rock boring (Fig. 3.24b) organisms (bivalves and gastropods) on the shore cliffs and shore platforms. They penetrate into the substances of rocks and woods for their security and protection in the marine intertidal environment. Rocks and woods become pitted and weakened by such boring activities of organisms which made them susceptible to erosion by other means of energy processes (waves, currents, winds, marine corrosion, etc.) in the region.

The terrestrial boring organisms, e.g. fungal elements of lichens have been acted as marine bores, especially on the cliff walls of the sea face around intertidal zones, which has led to the formation of undercut notches along the rocky shores of the islands. There is the presence of such significant features of undercut notches on the limestone boulders and large size drifted coral boulders over the shore platforms of Havelock and Little Andaman Islands. Undercut notches of Kalapathhar headland (Little Andaman) and Lakshmanpur headland (Neil Island) have been evolved into large scale development of notches by concentrated wave energies in every high tides situation (Fig. 3.24c, d).

The boring algae on the rocky shores can be grouped under epiliths, endoliths and chasmoliths on the basis of their attachment locations (Fig. 3.24e, f). Epiliths are usually associated with rock surfaces as they live on the upper surface of rocks along the shoreline. Endoliths live on the rocks below the top layer of epiliths, and behave like sponge of moistures to make the rocks to be weakened. Chasmoliths on the other-hand occupy the interstitial spaces between the rock grains and act as

Fig. 3.20 A sea cave, talus cave, taffoni marks and inland cave

Fig. 3.21 An underground cave (Endokarst) and the surface form of doline depression (Epikarst) in Kalapathar areas of Little Andaman

Table 3.5 Diversity of karstic and pseudokarstic caves of the South Andaman Islands identified in the field survey (Little Andaman)

Little Andaman					
Location	No of Caves	Latitude	Longitude	Elevation (m)	Rock types
A. *Hybrid Cave (Intertidal Coastal Zone)*					
Jackson Creek	2	10° 45′ 40.77″ N	92° 22′ 55.60″ E	11	Black Shale Limestone
		10° 43′ 14.01″ N	92° 22′ 59.81″ E	17	Black Shale Limestone
West Bay	3	10° 39′ 6.97″ N	92° 24′ 21.56″ E	9	Black Shale Limestone
		10° 36′ 20.65″ N	92° 24′ 46.53″ E	6	Black Shale Limestone
		10° 34′ 1.75″ N	92° 23′ 54.61″ E	8	Black Shale Limestone
South Bay	5	10° 31′ 6.90″ N	92° 28′ 51.30″ E	12	Black Shale Limestone
		10° 30′ 57.92″ N	92° 29′ 34.39″ E	18	Black Shale Limestone
		10° 30′ 40.77″ N	92° 30′ 17.52″ E	7	Black Shale Limestone
		10° 30′ 45.75″ N	92° 30′ 53.80″ E	13	Black Shale Limestone
		10° 31′ 2.31″ N	92° 31′ 25.71″ E	24	Black Shale Limestone
Hut Bay	2	10° 34′ 57.10″ N	92° 33′ 58.55″ E	11	Black Shale Limestone
		10° 35′ 28.96″ N	92° 32′ 41.29″ E	7	Black Shale Limestone
Kalapathar	4	10° 39′ 30.34″ N	92° 34′ 52.95″ E	11	Black Shale Limestone
		10° 41′ 33.11″ N	92° 35′ 45.29″ E	19	Black Shale Limestone
		10° 42′ 36.78″ N	92° 35′ 33.44″ E	17	Black Shale Limestone
		10° 44′ 53.43″ N	92° 34′ 56.69″ E	27	Black Shale Limestone
Dugong Creek	2	10° 49′ 38.89″ N	92° 33′ 47.49″ E	4	Black Shale Limestone
		10° 50′ 21.63″ N	92° 33′ 30.03″ E	14	Black Shale Limestone
B. *Flank Margin Caves (Inland from the Coast)*					

(continued)

Table 3.5 (continued)

Little Andaman					
Location	No of Caves	Latitude	Longitude	Elevation (m)	Rock types
Near Indira Bazar	3	10° 35′ 39.76″ N	92° 31′ 44.11″ E	25	Melville Limestone
		10° 34′ 6.45″ N	92° 30′ 30.47″ E	75	Melville Limestone
		10° 33′ 40.20″ N	92° 26′ 22.94″ E	61	Melville Limestone
Netaji Nagar	2	10° 40′ 49.29″ N	92° 34′ 44.89″ E	24	Melville Limestone
		10° 40′ 20.28″ N	92° 32′ 18.48″ E	79	Melville Limestone
Ramkrishnapur	2	10° 41′ 32.69″ N	92° 31′ 5.49″ E	88	Melville Limestone
		10° 43′ 14.98″ N	92° 29′ 18.64″ E	92	Melville Limestone
Whisper Waterfalls	1	10° 42′ 43.37″ N	92° 31′ 31.87″ E	65	Melville Limestone
Western part of Little Andaman	4	10° 44′ 56.94″ N	92° 25′ 44.09″ E	15	Melville Limestone
		10° 42′ 48.22″ N	92° 24′ 6.24″ E	80	Melville Limestone
		10° 40′ 51.04″ N	92° 26′ 14.87″ E	40	Melville Limestone
		10° 38′ 23.21″ N	92° 26′ 1.55″ E	79	Melville Limestone
Upper part of South Bay	1	10° 33′ 50.99″ N	92° 26′ 11.68″ E	51	Melville Limestone

an important agent of boring action into the rocks. The shore platforms adjacent of the headland of Lakshmanpur section (Neil Island) exhibit the formation of fretted surface due to the presence of endolithic perfarant algae.

Boring bivalves and boring barnacles penetrate into the carbonated rocks and also into the limestones very easily up to a few centimetres by producing tabular borings. The rocks of white clay stones, calcareous sandstones, mudstones, conglomerates and chalky limestones are mostly affected by such boring bivalve molluscs on the sea cliffs, stacks, fringe coral banks and shore platforms of the islands of South Andaman. They also produced tabular borings into the massive coral boulders drifted and attached on the shore platforms of Havelock Island and Neil Island. The living corals of Govind Nagar and Bijoy Nagar shore platforms exhibited the large colony of bivalve molluscs into the intertidal zone in which they have produced multiple boring sites up to 12 cm deep. The mangrove woods, peaty soils and vast open sand

Table 3.6 Karstic and pseudokarstic cave types of the South Andaman Islands identified in the field survey (Neil Island)

Neil Island					
Location	No of caves	Latitude	Longitude	Elevation (m)	Rock types
A. *Hybrid Cave (Intertidal Coastal Zone)*					
Lakshmanpur Beach-I	10	11° 49′ 50.92″ N	93° 1′ 0.10″ E	3	Sandy Limestone
		11° 49′ 47.96″ N	93° 0′ 58.09″ E	3	Sandy Limestone
		11° 49′ 45.08″ N	93° 0′ 56.46″ E	3	Sandy Limestone
		11° 49′ 47.54″ N	93° 0′ 55.25″ E	3	Sandy Limestone
		11° 49′ 49.37″ N	93° 0′ 53.67″ E	3	Sandy Limestone
		11° 49′ 53.88″ N	93° 0′ 51.08″ E	3	Sandy Limestone
		11° 49′ 57.66″ N	93° 0′ 52.72″ E	3	Sandy Limestone
		11° 50′ 30.21″ N	93° 0′ 47.15″ E	3	Sandy Limestone
		11° 50′ 27.74″ N	93° 4′ 6.22″ E	8	White Clay stone
		11° 50′ 11.25″ N	93° 4′ 14.93″ E	8	White Clay stone
Sitapur Beach	4	11° 49′ 54.17″ N	93° 4′ 10.20″ E	8	White Clay stone
		11°49′ 48.39″ N	93° 4′ 6.03″ E	8	White Clay stone
		11°49′ 40.66″ N	93° 4′ 1.24″ E	8	White Clay stone
		11°49′ 27.58″ N	93° 3′ 48.79″ E	8	White Clay stone
B. *Flank Margin Caves (Inland from the Coast)*					
Middle Part of the Neil Island	5	11° 49′ 26.12″ N	93° 3′ 34.72″ E	30	Clay Stone
		11° 49′ 43.69″ N	93° 2′ 20.83″ E	30	Clay Stone
		11° 49′ 51.13″ N	93° 2′ 48.88″ E	30	Clay Stone
		11° 50′ 6.61″ N	93° 2′ 52.15″ E	30	Clay Stone
		11° 50′ 15.19″ N	93° 3′ 10.95″ E	30	Clay Stone

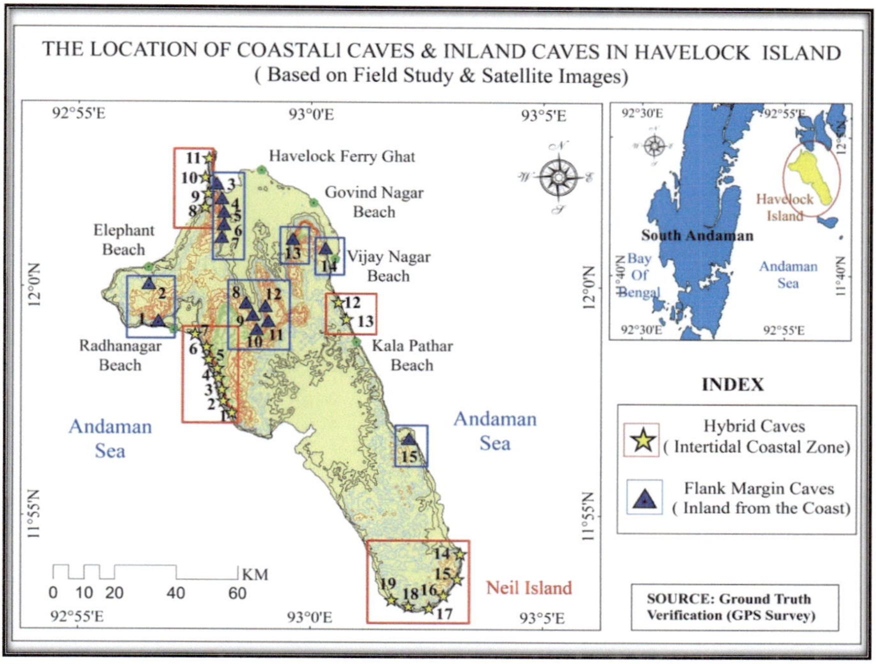

Fig. 3.22 The location map of karstic and pseudokarstic caves in Havelock

flats of the intertidal region are significantly affected and modified by activities of boring bivalve molluscs and boring barnacles in the tropical island shores of South Andaman. The group of *Lithophaga bivalve molluscs* and *Lithotrya barnacles* are commonest boring organisms in the tidal environment of the region. The large number of tide pools on the sand flats and rocky parts of the shore platforms represent diverse attachment colony of vermetid shell ridges in the islands (Fig. 3.24).

3.4.3.3 Other Phytokarstic Processes

A large part of near horizontal shore platform on the sea ward edge is thickly encrusted with Lithothamnion algae (brown colour algae) on the windward side of Neil Island (Fig. 3.25). The calcium cemented surface is wave beaten and flattened during rising tide levels, but evaporation in emergence (low tide exposures) gives rise to develop shrinkage cracks which attract the accumulation of calcium carbonates along the opening space of joints and develop linear ridges in rectangular forms.

Gradually, the jointed blocks with encrusted Lithothamnion algae provide conditions for development of chemical weathering, as well as water level weathering, and wave abrasion at the high tide levels, rendering to be weakened and eroded in the high wave energy environment. The calcification rate in the encrusted site is higher than

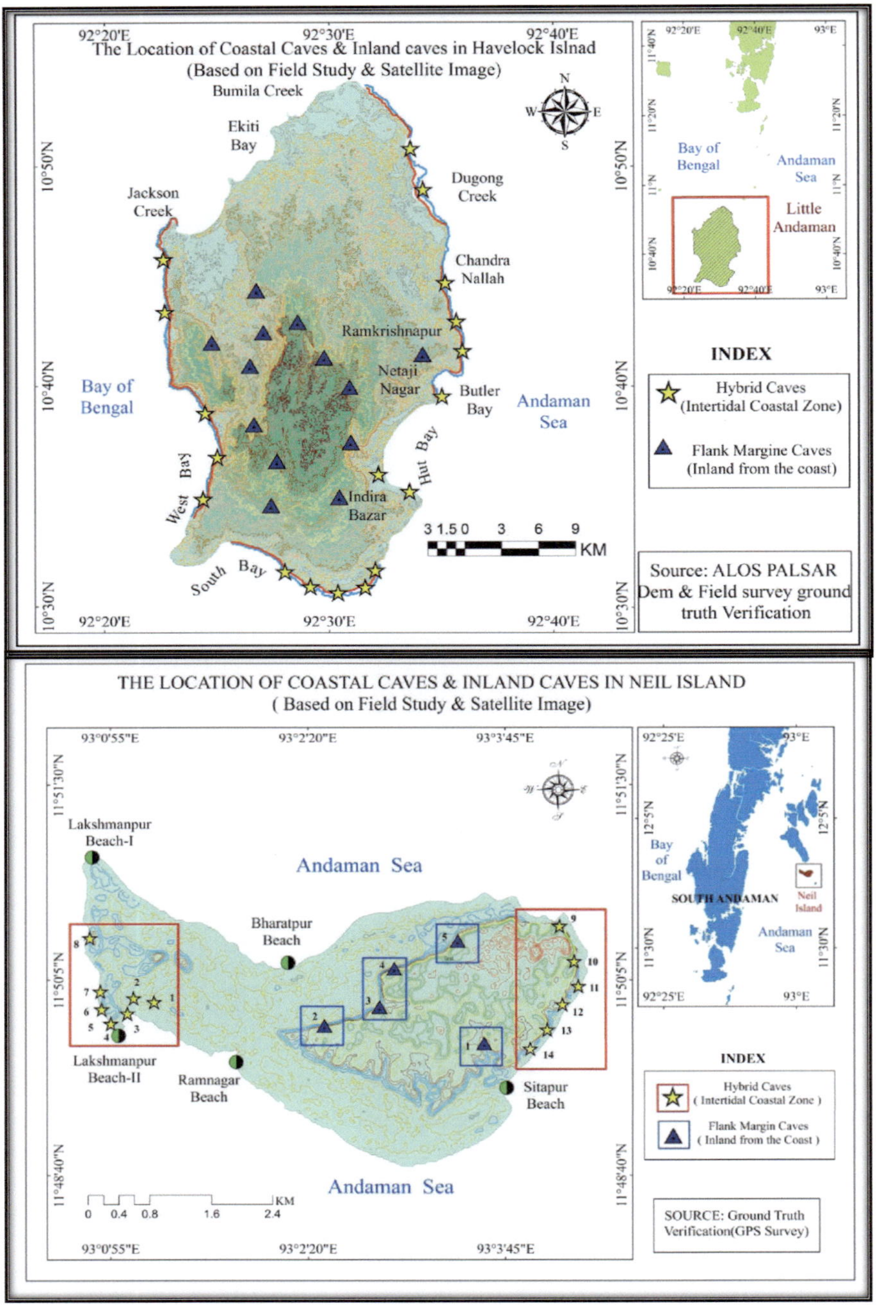

Fig. 3.23 Location of coastal caves and Inland caves in Little Andaman and Neil Island

Fig. 3.24 Various shore front Bio-karst Features of the South Andaman Islands; **a** Pinnacle Karst; **b** Oyster bank; **c** Rock boring bivalves; **d** Vermetid shell reef; **e** Boring algae and **f** Rock boring shells

Fig. 3.25 Degraded shallow water reefs (after 2004 Tsunami) with reduction of Zooxanthellae in Neil Island (Lithothamnian Flat)

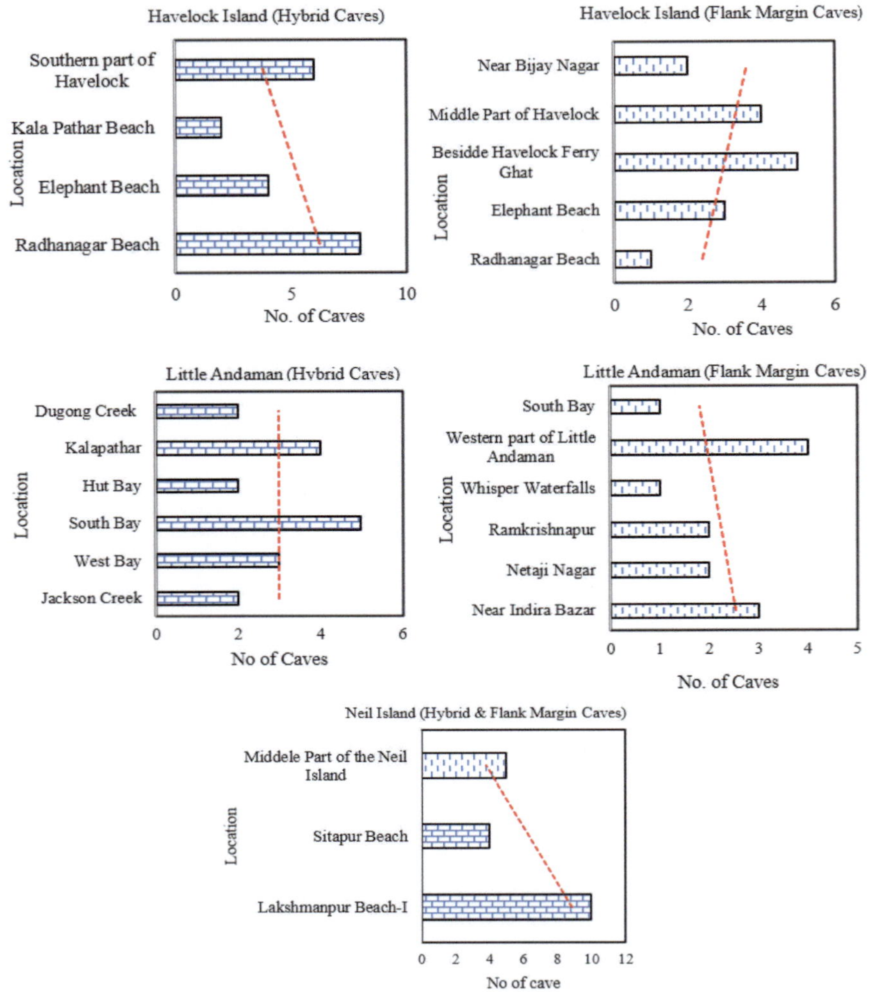

Fig. 3.26 Frequency of cave distribution in the form of hybrid caves and Flank Margin caves

the open shore platform to develop the phytokarstic features in the Lithothamnion flat of the environment (Fig. 3.25).

Sea grasses colony development over the rock pools not only increase the rate of calcification (10 kg/m^2 per year) but also produce patches of coastal habitats in a smaller pockets on the inter tidal shore platforms. On evaporation, salt weathering becomes an active process in the pool surface to disintegrate the weaker rocks in the shore platforms. However, the full grown weeds or sea grasses, their stems, leaves and rhizomes may be torn away by storm waves or by swell waves and washed on to the shore platforms of the region. As the rotting sea grass produced acidic water in

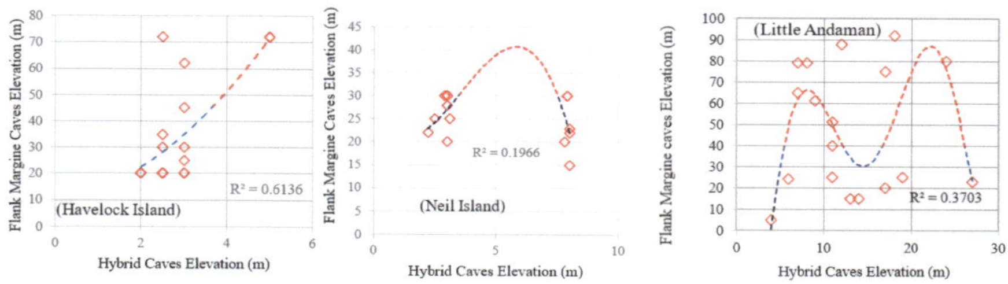

Fig. 3.27 Regression analysis on the elevation differences between hybrid caves and Flank margin caves showing the highly positive relationship particularly in Havelock Island

the rock pools that dissolved the shelly sands and removed iron oxide coatings from sand grains in the areas of beach rocks (Lakshmanpur point, Neil Island).

Oyster reefs are built on the strong attachment devices, e.g. boulders, steeper banks and rocky shores of the island fringe tidal and sub tidal regions. The oysters bore into the rocks and produce reefs or minor ridges on the attachment devices in the tropical coast. There are number of drifted boulders lying on the shore platforms in Havelock Island and in parts of Neil Island after the events of large-scale marine inundations (2004, 2016, etc.), exhibited attachment devices with oyster reefs in the region. Such evidence proves that they are undermined and drifted from the reef front bank margins by strong wave energy with partial attachment surface of oysters (Fig. 3.24e).

Vermetid reefs are shell ridges usually found on the pools of water bodies over the rocky shores (shore platforms) on the compact sand flats, and also around the swampy tract of mangroves under carbonate settings in the island shores. The ridges accumulated carbonate sediments and reduced the wave energy in places of the shore platforms. Such resistant reefs are also found on the wood logs and jointed rocks of the sea shores (Fig. 3.24d).

Encrusting corals (Figs. 3.24 and 3.29) are low spreading corals grown the flat rocky surfaces in attached forms in the intertidal shore platforms of all three islands of South Andaman. They mostly belong to Montipora and Porites group and usually adhere to hard rocky surfaces to resist wave energy along the extensive pools of water bodies on the shore platforms. Closely cemented shore platforms with coral rags and blackened algae mats over the limestone rocks reduces wave energy at present but represented grooves and spurs of older origin at the lower part of cemented rags. The relict signatures of spurs and grooves were produced in a paleo reef front margins which has emerged above the mean sea level during seismo tectonic events. Such fretted surface with reef rocks are affected by weathering and erosion during seepage flow of fresh waters from the landward sides and by wave abrasion during high water levels in the region of Neil Island (wind ward side) adjacent of Ramnagar section (Fig. 3.28).

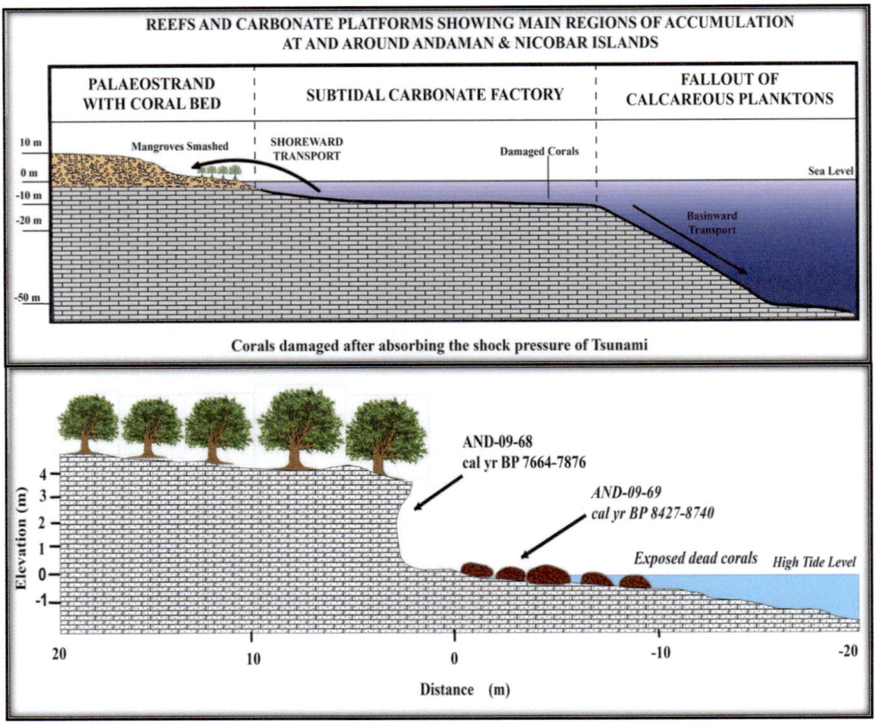

Fig. 3.28 The shore platform sediments of Neil Island (above) and boulder drifts in Havelock Island (below) after the impact of Great Earthquake 2004

Fig. 3.29 Surface encrusted corals

3.5 Beach Rock Formation and Materials

The well cemented sedimentary rocks with carbonate minerals along the shoreline of different grains of sands, shingles, gravels and finer materials are usually visible on the intertidal zones of tropical islands. They are known as beach rocks and widely distributed along the upper part of intertidal zones particularly of the embayed shores in Neil Island.

3.5.1 Beach Rocks and Karstification

The precipitation of carbonated in the intertidal zone of fluctuating water table with rise and fall of tides and seasonal alternations of wetness (May–October) and dryness (November to April) can accumulate and cement the beach sands into sandstones known as beach rocks in the island shores (Bird, 2008). Beach rocks of calcareous sandstones (or beach calcarenites) have been found on the shores of Neil Island particularly on the windward sides around the coasts of Andaman Sea. The vertical movement of water table of the seashores is closely related with the tidal pumping at regular interval. During high tides the water table goes down by pressure of tides, and upward movement of water table takes place at the low tide exposures along the shoreline which causes the movement of dissolved carbonates into the beach sands of the shore platforms in the island (Fig. 3.30). The beach sands were consolidated with the addition of calcium carbonates or the calcites, aragonites, dolomites, etc., cementing agents precipitated or mixed with the beach materials in the intertidal zone. Shore platforms of Neil Island are attached with the algal encrustation, which inhabit the shores close to the water table, and provided conditions for precipitation of calcium carbonates to consolidate the sandy and clayey materials in the tidal zones (Figs. 3.29 and 3.30).

The tropical warm and humid climates with strong evaporation rate, high temperature and high interstitial water temperature (over 25 °C) for most of the year supported the environment for rapid cementation of beach sands and other bioclasts in the region. The surface morphometry of the extensive calcareous sandy beach rocks (Lakshmanpur section) exhibit series of shore parallel clifflets with dip and antidip slopes, folded, faulted layers, segmented ridges and sharp-edged karren features. Weathering, wave abrasion and denudational processes have modified the hard sandstones into erosive section of the embayed shores (Fig. 3.30).

Sand clay sediments mixed with graveliferous coral debris are also converted into beach rocks along the shorelines of Govindnagar section of Neil Island. They are partially exposed in the intertidal zone after erosion in this section of the shoreline in Andaman Sea. In both cases of the beach rock formation, low lying shores with seepage flow of ground water and availability of sand cay sediments are common. However, the overlying sands have occupied the landward extension of beach rocks in the region of Neil Island.

Fig. 3.30 Beach rocks with calcareous sandstone, graveliferous materials, beach sands and isolated ridge cusp materials in Neil Island

3.5.2 *Degassing of Carbonate Platforms*

As the underlying carbonate rocks of the coastal fringes environment become weathered after rain water percolation through the joints, the calcite precipitation takes place in the high tide zone through the seepage flow of ground waters. The precipitation of calcites into the tidal zones by the process is known as degassing of carbonates. The coastal fringe water table also fluctuates with vertical movement by tidal pumping in the shore platforms, and during low tide exposures as the pressure drops the calcites precipitate into the intertidal zones through seepage flow of dissolved carbonates (Fig. 3.31). Calcites precipitation in this process also known as degassing of carbonates and influence the cay sediments or the beach conglomerates in cementation of materials under warm environment of the tropics. Field studies reveal that water seeps through the cay sediments along the high tide line during emergence of the shore platforms in low tide condition of pre-monsoon months (Feb–March).

Fig. 3.31 Degassing of calcium carbonates by seepage flow from the shore fringe area at the HTL

3.5.3 Chemical Processes of Landforms

Chemical weathering with solution process, carbonation process, water layer weathering and salt weathering processes on the calcareous beach rock sandstones of Neil Island displayed a distinctive topography with pits, linear pans, micro pits and spikes. Boring algaes (blue-green colour cyanobacteria) also helped in the formation of such coastal karren, with the action of sea waves, wetting and drying, and hydration process on the intertidal shore platforms of the island coast (Figs. 3.31 and 3.32).

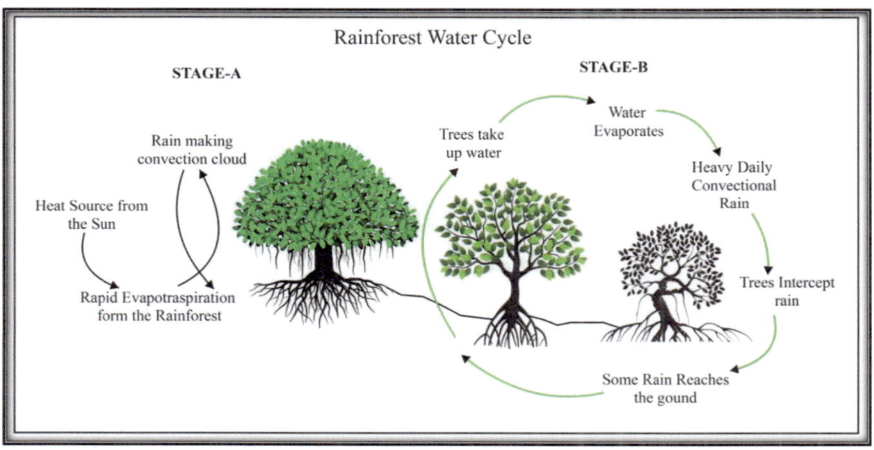

Fig. 3.32 The rainforest water cycle played significant role for the existence of biological processes in weathering of rocks

3.6 Coastal Hydraulics and Karstic Landforms

Hydrological processes of the islands including with shallow sea waves, winds, currents, tides, marine inundations and tsunamis played a major role in modification of landforms, and acting as the signatures or impacts of coastal hydraulics. The landforms and their causative processes are discussed in this section particularly on the sea face considering with shoreline configurations and shore cliff structures of the islands.

3.6.1 Island Hydrological Cycle

The island systems of the Andaman Sea and the Bay of Bengal (Havelock, Neil and Little Andaman) maintain their self-sustaining hydrological cycle with developed rain forest ecosystem. The role of rain forests in water cycle is very significant as the water released through transpiration process by plant leaves supply water or moisture into the atmosphere during photosynthesis in the region of tropical warm climate (Fig. 3.32). About 99% areas of Little Andaman are densely covered by canopy layers of tropical rainforests at present, and these vegetations contribute significant level of moistures to the formation of rain clouds, which usually release water back onto the rainforests of the island fringed with the Bay of Bengal and Andaman Sea.

The typical convectional rainfall is guided by the temperatures (30–32 °C) of such tropical islands, and when the canopy heats up during day time, the water evaporates into the atmosphere to form clouds over the region for next day's rain. The deeper roots and sub-surface roots of tropical forests take up water from the ground layers and soil layers, during day time and rain waters of significant volume are also intercepted in their canopy levels but they release water into the air by evapotranspiration rates for the formation of clouds and convectional rains.

Although islands of the regional setting are characterised by the occurrences of convectional rains, monsoonal rains, and cyclonic rains, the local rainfall is entirely guided by the cover of forest lands. Little Andaman receives about 4000 mm of annual rain, but Havelock and Neil Islands receive 3500 mm of annual rain due to the presence of large forest cover areas of Little Andaman in compare to the other islands of South Andaman district.

3.6.2 Waves, Currents and Tides

On the basis of height and energy levels, the sea waves can be categorised into wind waves, swell waves, storm surge waves and tsunami waves in the region of coastal and open marine environment. The INCOIS real time data displayed the nature of waves, winds, currents and tides of the region of Andaman Sea and adjacent areas of the Bay

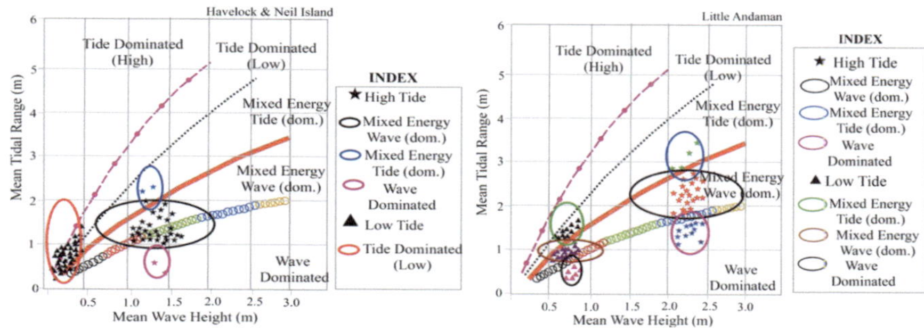

Fig. 3.33 Various process dominant coastal margins of the islands

of Bengal. Havelock Island and Neil Island of South Andaman exhibited the wave heights ranging from 1.5 to 1.8 m in the month of April, 2021 and Little Andaman Island of main Andaman Ridge fringes with the Bay of Bengal and Andaman Sea represented 2.0–2.50 m wave height during the same period in the region (Fig. 3.33).

The occurrences of monsoon drift currents approached the shore fringes of the islands from north-east and south-west directions in two different seasons (winter and summer monsoons). Islands of the region also registered erosional and depositional features on the basis of their location and shoreline configurations by approaching towards the windward and leeward sides of the drift currents. Tidal amplitudes are relatively lower than the wave heights in the region of Andaman Sea as recorded in Port Blair and Hut Bay Stations (Figs. 3.33 and 3.34).

Tidal range varies from 0.70 to 0.38 m in Havelock and Neil Island, and it represents 1.45–0.80 m in Little Andaman. Tidal elevations and encroachments push or advance the wave breakers into the landward fringes of the island shores (Fig. 3.34).

Andaman Islands are largely affected by west coast swell waves and east coast swell waves as per the location of Andaman ridge extended from north to south. The long distant waves guided by west winds and trade winds provide wave energy amplification and approach the shore fringe areas of carbonate islands with marine inundations. Coral reef growths on the west coast are healthier than the east coast of Andaman ridge due to energy variation of swell waves.

Storm surge waves and Tsunami waves generated large-scale marine inundations in the coastal tracts fringed with shore platforms and low-lying terraces. The variation of sediment deposition and transportation is influenced along the island shore by such events of large-scale marine waves in the past.

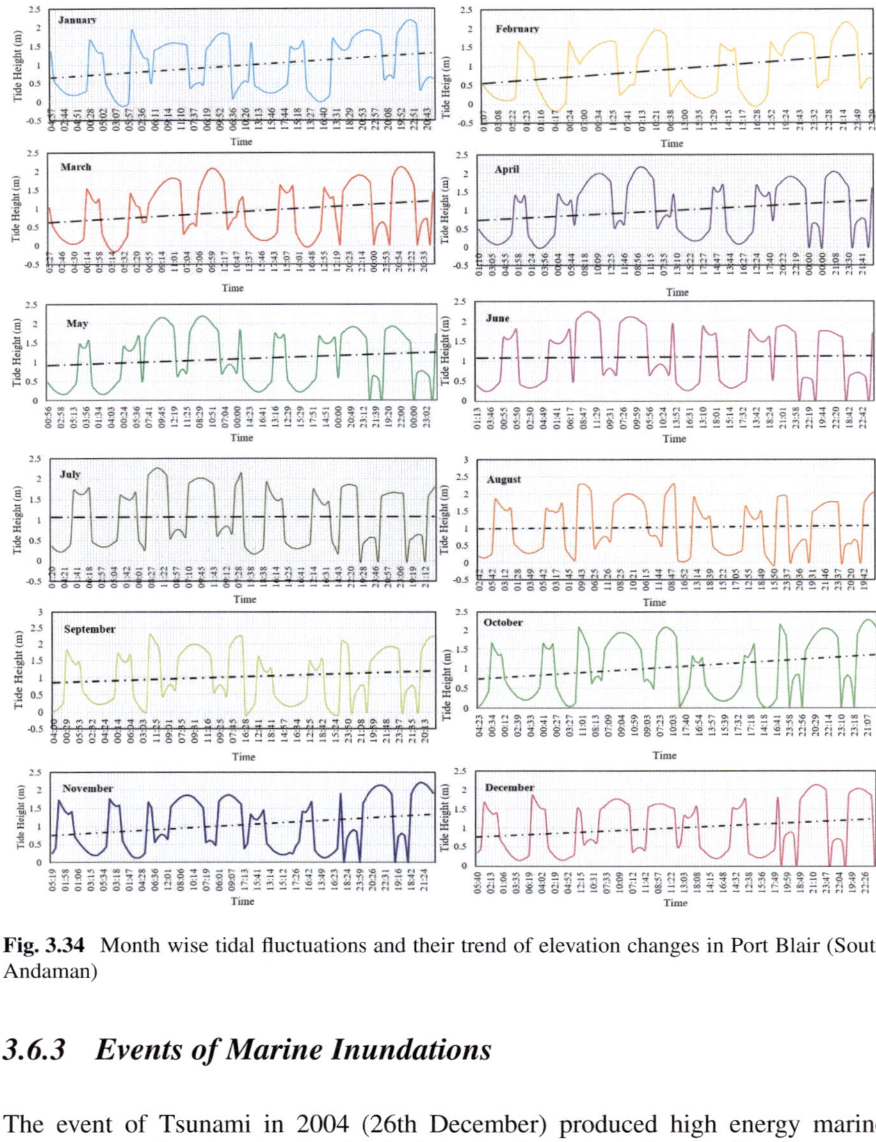

Fig. 3.34 Month wise tidal fluctuations and their trend of elevation changes in Port Blair (South Andaman)

3.6.3 Events of Marine Inundations

The event of Tsunami in 2004 (26th December) produced high energy marine waves up to 10 m in elevation in the islands of South Andaman after the occurrences of sea bed earthquake (north-western offshores of Sumatra Island) with 9.1 Richter Scale magnitude. Such tsunami waves overturned coral heads and destroyed most branching forms of Acropora corals across the island fringe location (Paul, 2005). Spur and groove formations of the steeper banks in the fore reef environment disrupted by high energy tsunami waves. The existing patch reefs of the intertidal shore platforms were blanketed by film of sediment layers in the event of marine inundation. Large boulders were dislodged from the oyster bank, reef front bank,

Fig. 3.35 Wash over transport of graveliferous sediments by the marine inundations with cyclonic surge waves

and fringe coral reef complex by the bottom friction with advanced tsunami waves, and they were transported towards the shore platform surfaces of the islands after the event of marine inundations. Pebble and rubble ridges were formed along the baseline areas of shore cliffs on the windward sides of the islands (Fig. 3.35).

The events of marine inundation also took place in the island shores by cyclonic storm surge generated waves during 28.11.2013 (Cyclone Lehar) and 06.12.2016 (Cyclone Vardah). Such events of marine inundations transported inorganic sands and shingles into the shore platforms of Havelock and Neil Island.

Cay sediments were transported and deposited on the beaches and dunes above the shore platforms by such storm surge waves in these carbonate islands. The over wash transport of gravels, shingles and sands have built up terraces with beach calcarenites in the region after the events of marine inundations up to 5 m in elevation.

3.6.4 Boulder Size Zingg's Classification

The sections of shore platforms are selected in Havelock and Neil Island for sampling the pebbles and boulder size sediments in different period of field visits (2012–2018). Number of white clay stone debris and coral debris were measured in Kalapathhar point of Havelock Island with the attributes of length (L), width (W) and thickness (T). Shape and size analysis of the sediment samples are followed by Zingg's Classification (Fig. 3.36).

Shore platforms of Lakshmanpur section (Neil) and Radhanagar West section (Havelock) accumulated pebbles, coral rubbles, and coral boulders on their windward sides. High energy waves modified the roughness of boulder surface. They are also attached with oysters and barnacles (Fig. 3.36).

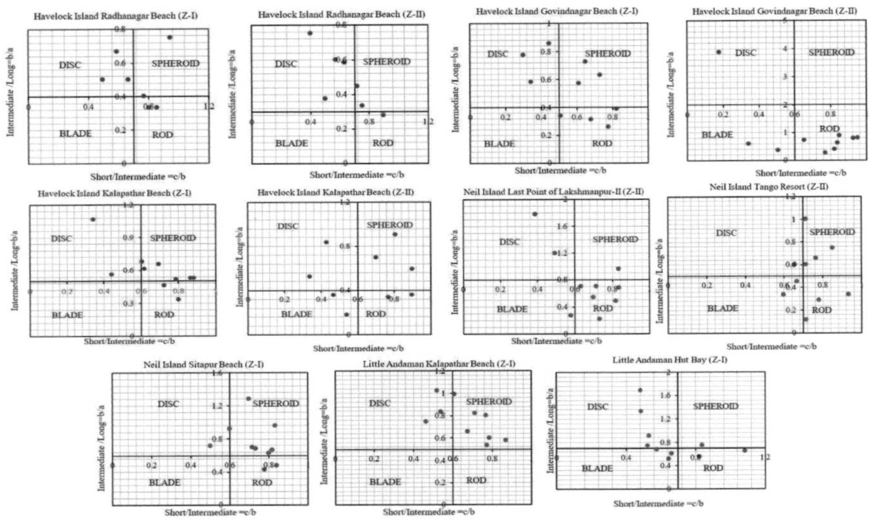

Fig. 3.36 Different types of the shapes of the boulder analysed by Zingg's method

Chapter 4
Reef Islands and Carbonate Sediments

4.1 Coral Reefs of the South Andaman Islands

The shallow marine environment of the Andaman and Nicobar Islands exhibits well-established fringing reefs on the east coast and barrier reefs on the west coast, channel reefs on the estuaries and straits, and the patch reefs into on the estuaries and straits, and the patch reefs into the water pockets of wider shore platforms. They occupy about 13,500 km^2 areas of the sea beds with diverse reef species, including hard corals and soft corals. Coral reefs of the archipelago are considered as biodiversity hotspot, and they act as a submarine bioshield, supporting huge marine life, a supplier of reef sediments, absorber of the carbon sink and other pollutants in the region of the tropical oceanic environment. Major reef species of the island system are Acropora, Pocillopora, Porites, Goniopora, Favia, Montipora, Millepora, Fungia, etc. There are 135 species of corals in the Andaman and Nicobar group of Islands (Paul and Paul, 2023a, 2023b, Pillai, 1983, Roy et al., 2014).

Some islands are notable for their reef systems of the region under the Andaman fringe sea and the Bay of Bengal Sea, such as North Bay, Neil Island, Havelock Island, Ross Island, Rutland Island, and Little Andaman Island, North Sentinel, Red Skin Island, Jolly Buoy Island, etc. Reef flats are marked by wider and extensive habitats on the West Coast of the Andaman group of Islands, whereas reef banks extend over a gently sloping and stepped surface across the shallow marine environment on the east coast of the mainland archipelago (Fig. 4.1). The fringing reefs of the Andaman Sea were badly affected by developed fissures, sedimentation, erosion and removal of algal covers on the reef banks during the 2004 great earthquake and tsunami events. However, the barrier reefs of the west coast were damaged due to the vertical changes of the shoreline in similar events (Fig. 4.1).

Reefs of the channels and straits are dominated by soft corals. The Duncan Passage Strait between Rutland Island and Chidiya Tapu supported very attractive and diverse corals in the sheltered habitat, but was affected by subsidence and siltation during the tsunami event. Baratang channel strait on the northern end of the South Andaman

A. Paul et al., *Karstified Coasts of the South Andaman Islands*,
https://doi.org/10.1007/978-3-032-10299-7_4

Fig. 4.1 The patch corals of the shore platform water pockets of South Andaman Islands. The coral reef niches were heavily damaged by the sedimentation and boulder drifting across the shore platforms during the earthquake and tsunami events (2004)

district exhibited very good habitat for soft corals and a few encrusted corals. The channels' fringes are lined with tidal woodlands of mangroves. The coral reefs of the channels and straits supported attractive marine life dominated by shells, sponges, weeds, algae and gastropods.

However, the watery basins, lagoons, tide pools or other water pockets of the shore platforms are dominated by colourful encrusted corals, patch corals and table corals. Such types of coral colonies are associated with weeds, seagrasses and algal encrustations. Marine animals such as sea cucumbers, sea anemones, sea urchins, ornamental fishes, sea stars of many species, barnacles, bivalves and encrustations are abundant in the patch coral habitat of the South Andaman Islands (Fig. 4.2a, b). During the events of earthquakes and tsunamis, the encrusted corals were affected by sedimentation and boulder drifts across the shore platforms. The biomass of photo synthesizers organisms is an important primary producer in the coral reef ecosystems under sheltered habitats.

The coral reefs of the Andaman and Nicobar Islands are sharply distributed in different niches and favourable topography defined in conditions of suitable currents, wave exposures, variations in salinity, water depth, attachment surfaces, turbidity, temperature and other factors. Coral growths are luxuriant in the windward sides

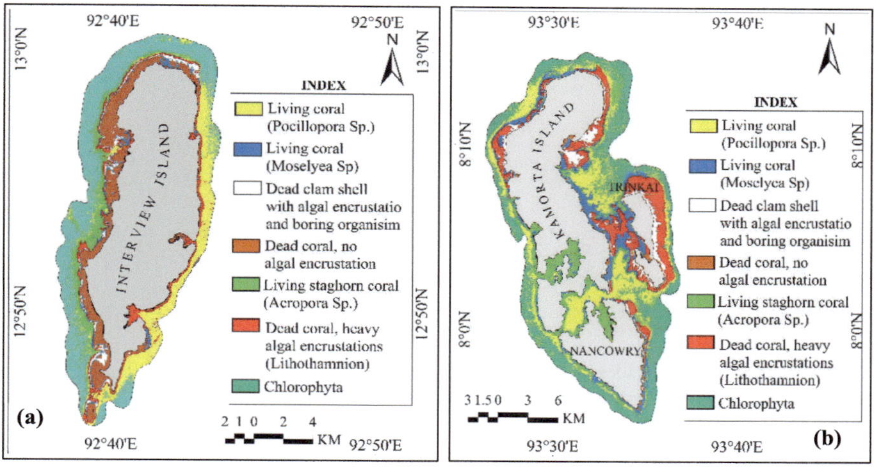

Fig. 4.2 The Landsat 8 OLI-based coral reef classification techniques using the infrared band **a** fringing coral reef in the North Sentinel Island of the Bay of Bengal; and **b** fringing coral reefs of Nicobar group of islands system (Kamorta, Nancowry and Trinkat islands)

of the tropical oceanic islands (Fig. 4.3). Cay sediments have redefined the shoreline positions of the leeward sides after the tsunami surge waves in the island systems. Debris of the hard corals such as acropora, montipora, favia and porites was distributed on the shorelines in the form of rubble and boulders after the events of earthquakes and tsunami waves in the island fringes. The Holocene reefs and the reef systems of LGM (400 YBP to 21,000 YBP) were affected by seismo-tectonics and converted into raised reef terraces and submerged reefs with fluctuating sea levels, as well as by changing ecological niches and topography. Pleistocene reef limestones are lying at the basement below the Holocene reef systems, but island fringes are affected by wave cut platforms with oscillations of sea levels in the region. However, reef terraces over the shore platforms are wave-abraded and sharply eroded, which turned into the residual benches in the South Andaman Islands.

4.2 Reef Sedimentary Landforms

Presently, the reef sedimentary materials such as reef rubbles, reef boulders or debris, branching coral fragments, coralline sands, stony corals, beach rocks and other bioclasts are deposited and lithified on the older reef platforms of the limestone pavements along the shore fringe environments of the islands (Fig. 4.4a). The calcium-rich beach sands with foraminifers are lithified and cemented as beach rocks in the high tide levels and above the high tide levels during the Holocene epoch, and they are also

Fig. 4.3 Patch coral reef under the development of the shore platform in the Neil Island (*Psammacora obtusangura*). The encrusted coral is affected by the sedimentation during the great earthquake events

called as beach calcarenites. However, the island platforms in many places constitute carbonate sands which blew onshore by winds during the last glacial maximum (Late Pleistocene epoch, 18,000–21,000 YBP) and dunes, beaches and bars have emerged with unconsolidated sediments with foraminifers at the supratidal zones, which turned to rocks after lithification process over the time. Thus, the present-day raised beach areas of the island platforms contain eolinites and beach rocks. Probably they were reworked by the wind waves during the Holocene transgression in the region, where the carbonate sediments are mixed with coral fragments and marine shells (Fig. 4.4b).

The reef banks and reef flats are fringed by coral colonies in which broken corals, sand grains, the skeleton of crustaceans, weeds and algal materials, as well as marine shell fragments mixed with reef boring animals' bodies prevail, which turned into reef rocks by cementation (Fig. 4.4c). The reef zones supply carbonate sediments into the shore platforms during surge waves, surge tides and tsunami waves to run the morphodynamics of the shore fringe environment. Carbonate sediments are also supplied into the shore platforms in a solution process during monsoon months from the island platforms after rains. The mass wasting materials after the cliff failures (Benumof et al., 2000), a large amount of debris, boulders and earth materials at the foot slopes, become reworked by advancing waves during high tides and contribute some amount of carbonate sediments into the shore platforms of the islands (Fig. 4.4d).

Reef sedimentary landforms are diverse in character and they extend from the shallow marine environment as reef flats, reef rim, reef bank, reef front, reef slope, reef platforms with patch reef, reef rubbles, reef boulders or debris, carbonate sands, broken reefs, coralline sands, beach rocks, raised beach surfaces and reef terraces up to the island platforms. Carbonate sediment-filled rai shores, sediment cays, sand bars and sand ridges are developed along the redefined shores after the 2004 tsunami waves in the islands.

Fig. 4.4 Reef sediments supplied to the shore platform and associated landforms of Neil Island. **a** Carbonated tidal flats are heavily bioturbated, **b** limestone boulders are drifted into the floor of the frontal mangroves niche, **c** coral fragments of *Acropora Sp.* and debris corals of *Montipora Sp.* drifted into the sinks of the shore platforms and **d** coral rubbles with boulders drifted and were affected by solution pits

4.3 The Reefs as Morphodynamic System

The health of the coral reefs displays the actual structure and function of the ecosystem. As the physical and biological processes are constantly interacting, shaping their morphology and the surrounding environment, the coral reefs are considered as dynamic ecosystems. Thus, the complex interaction and interplay of multiple factors represent the coral reefs as a morphodynamics system. The Andaman and Nicobar Islands are fringed by coral reefs in the tropical oceanic environment, extending from 6°45' North latitude to 13°15' North latitude. For the activities of west coast swell waves, the coral reefs and their growths are luxuriant on the Bay of Bengal fringe coast of the archipelago. Quality of the reefs varies at different depths, ranging from 2 to 15 m, 15 m to 30 m and from 30 to 50 m depth along the banks of the underwater Andaman ridge. The archipelago is not continuous but broken in places by the location of straits and wider channels connecting the Andaman Sea with the Bay of Bengal from east to west. Many sub-habitats of the coral reefs are identified in the Bay of Bengal and the Andaman Sea fringes of the shallow marine environment. Physiographically, they are categorised as (i) reefs banks, (ii) reef flats, (iii) reef fronts, (iv) strait margins, (v) estuary fringes, and (vi) shore platforms in the island systems. Coral reef problems are, however, very complex as they have to interact, interplay and adjust with the stresses produced by multiple factors of physical and biological processes in the marine environment. Major physical factors include sea level oscillations, vertical deformations by seismo-tectonics, tsunami hazards, attachment surface, weathering rates, erosion rates and natural growth rates of corals. Among the biological factors, the competition for space, diseases, predation and the impact of other organisms, coral boring species and human impacts are significant (Fig. 4.5a).

Thus, interaction and interplay between the physical and biological factors with the coral reefs displayed the present-day morphodynamics system of the shallow marine habitat at the Andaman and Nicobar group of islands (Fig. 4.5c, d). The effects of seismotectonic, tsunami events, predation and high SST were reflected over the coral reef habitat in the previous decades (1980–2020s) in the oceanic islands. Coral reefs in many places of the archipelago have been raised, drowned, bleached, fractured, affected by erosion and sedimentation during a tsunami event,

Fig. 4.5 Older reefs of the shore platforms eroded and wave abraded in the windward sides of the South Andaman Islands. **a** Beach calcarenites lithified and cemented as beach rocks in the high tidal zones, **b** wave-cut notches at the foot slope of the limestone cliff along the shoreline, **c** wave-abraded coastal karrens on the edge of the shore platforms and **d** lithothamnion flat case hardened and cemented over the coralline algal ridges of the shore platforms at the reef front position

and significantly predated. The morphodynamic adjustment of coral reefs is taking place to maintain the damaged habitats in normal conditions, the natural functions of the shallow marine bio shield, and the rich biodiversity in the oceanic environment. Coral reefs of the Havelock Island, Jolly Buoy Island, North Bay and Ross Island were severely affected by the removal of Zooxanthella cover from the reef habitat during the tsunami event. Patch reefs of the shore platforms of the south Andaman Islands were declined and damaged by the cover of sediment layers deposited in a similar marine hazard (2004). Thus, the feedback mechanisms of the coral reef morphodynamics show negative and positive changes of the habitat in the previous decades (Fig. 4.5b).

4.4 Carbonate Factory and Environmental Factors

Production of carbonate sediments occurs in the fringing reef systems of the islands under a favourable shallow marine environment. The older reefs with shore platforms and island platforms also supply carbonate sediments under the activities of weathering, sub-aerial exposures and erosion processes. Marine biogenic calcification, dissolved compounds of the sea waters, ions, nutrients and the degree of saturated sea waters provide the sources of $CaCO_3$, which contributes a major role in reef-building mechanisms in the shallow marine environment. Attachment surfaces, penetration of sunlight, cover of zooxanthellae, photosynthesis power and coral polyps also jointly produce conditions for of reef colony.

Under such favourable conditions of shallow marine environment of the tropical oceanic islands, the coral colonies accumulated broken coral species, fragments of marine shells, biogenic erosion of primary reefs by rock borers, sea weeds and other bio-clasts, and calcium carbonates which tends to become carbonate sediments in loose condition but converted into reef rocks by lithification and cementation (Figs. 4.6, 4.7a).

The loose carbonate sediments after production may be transported to the shore platforms or partially to the pelagic environment by storm surges, wind waves or by tsunami waves as organic materials into the sub-systems (Fig. 4.7b). Thus, the shore fringe platforms of the islands accumulated the carbonated sediments as cover deposits by sea waves and tidal waves and took part in the shore platforms' morphodynamics. Longshore currents and cross currents transport and distribute the accumulated carbonate sediments into the favourable sinks of the shore platforms on the basis of their topographic factors (Fig. 4.7ac, d).

The sea level oscillations, seismo-tectonics and other marine hazards acted as the energy drivers on the shore platforms to redistribute the carbonate sediments and modification of surface morphology. Eroded materials from older reefs and island platforms contributed additional carbonate sediments into the shore platforms. The cover deposits, pavement surface, beach rocks, boulder drifts, rock pools with sediments, beach deposits and wash over sediment lobes are the sinks of carbonate sediments in different forms across the shore platforms. Several micro habitats of

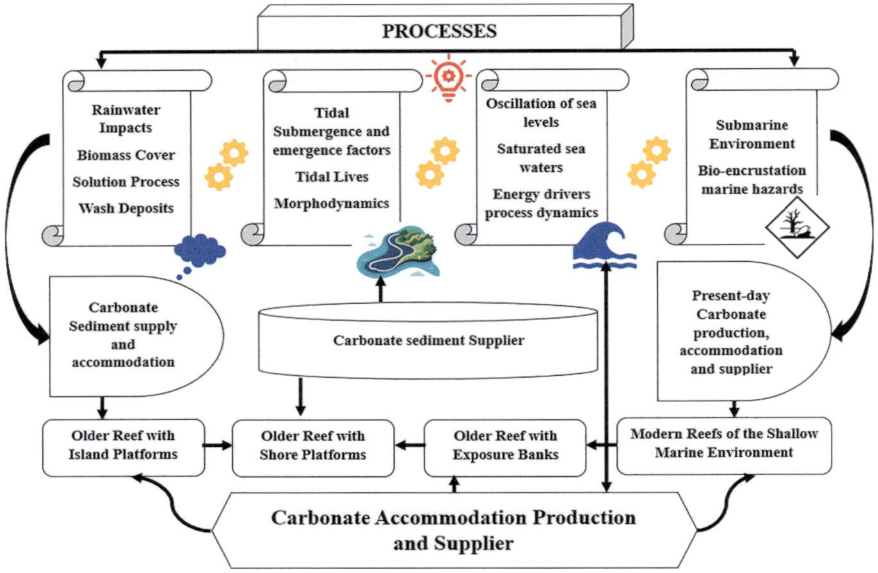

Fig. 4.6 The diagram shows the interaction and interplaying processes of the environmental factors for the carbonate machine in the shore fringe marine environment of Andaman and Nicobar Islands

the shore platforms, such as mangrove swamps, patch reefs, tide pools, algal mats or ridges and other forms of bio-encrustations, also produce carbonate sediments in the shore fringe environments (Fig. 4.7a, b). Under a stable condition of the shore platforms, the carbonate sediments of the high-tide shores were lithified and cemented to develop beach rocks, and limestone pavements associated with boulders and debris changed into Karstified forms by the solution process in the tropical coast.

4.5 Karstified Carbonate Platforms

Various karst features are developed in the ground surface of the carbonate shore platforms, vertical cliff walls of the joined limestones, promontories, headlands of jointed limestones and also visible at the frontal reef banks of the seaward edge of shore platforms of the Andaman Islands. The bicarbonate solution and dissolution processes, jointed and thicker limestones, other carbonate rocks such as calcareous sandstones and reef rocks, tropical rainfalls and temperatures have created very distinctive landforms of karstic features in the coastal fringes.

Along the frontal reef banks of the older reef rocks, several karstic features such as solution pits, solution pans, blowholes, tunnels and underwater caves were formed by weathering and erosion processes when the sea level was slightly lower than the

Fig. 4.7 Loose carbonate sediments drifted into the shore platforms by hydraulic energies. **a** Coral rudstone or floatstone deposited in the topographic depression of the shore platform, **b** the wave-driven energies through the geos played the role to transport the coralline debris, **c** fragmented limestone pavements drifted into the high tide zone marked with solution pits and lappies, and **d** boulder drifts across the shore platform dominated by frontal mangroves

present-day sea level. Presently, they are wave eroded, weathered and extended into distinct features at the south Andaman Islands (Fig. 4.8a, b, c, d, e, f).

Over the ground surface of the shore platforms, very distinct features of karstification are reflected in the carbonate rocks (Lace and Mylroie, 2013). Tafoni marks and rillen karren features on the limestone blocks, clints and grikes on the limestone pavements, pinnacle karst on the limestone ridges, solution spikes and coastal karren pitting on the bio-encrusted shore platforms, and Lapies on the reef limestone blocks are very common along the island fringing shore platforms (Fig. 4.8b, c, e). The shore platforms of South Andaman also represented the formation of limestone solution notches and doline depressions by piping, subsidence and surface solution processes.

Finally, the limestone ridges, shore cliffs, promontories and headlands of the island platforms are marked by the various karst features in the Andaman Islands. Tower karst, fissure caves, tunnels, cliff base notches, geos, dolines, seepage pits, tafoniferous walls and flanks margin sea caves are found in Little Andaman, Neil Island and Havelock Island of South Andaman (Fig. 4.8e, f). The residual hills of limestone rock are standing as towers with 40 m or more tall on the bank of the

Fig. 4.8 Distinctive karstified features of South Andaman Islands on the coastal fringes. **a** Wave-abraded potholes on the shore platform at the landward edge, **b** notches and fissure caves at the foot slope of the limestone shore cliffs, **c** prominent spikes of the coastal karren feature, **d** wave-abraded coastal karrens in the shore platform, **e** roof collapse in the caves developed into the chalky limestone cliffs of Sitapur and **f** abundant fissure caves of the limestone ridges at the edge of the promontory of Havelock Island

Kalapaththar landscape near Buttler Bay. The shale limestones of Little Andaman represented very distinct karst features in the South Andaman province.

4.6 Sediment Supply on Island Building

Island enlargement may occur with deposition of sand cay sediments, beach deposits and coral debris as well as by drifted limestone blocks on the island banks by tsunami waves, storm surge waves and swell waves in the tropical oceanic environment. Presently, the allochthonous sediments of medium grained sands are largely dumped into the shore platforms after the tsunami attack (2004) and cyclonic weather (2014), which played a significant role in damaging mangroves by trapping the breathing roots on the shore platforms. The supply of coral rubbles, branching coral fragments, coralline sands, beach ridge deposits and tsunami wave-driven sea bed sands has redefined the shorelines of the smaller islands of the South Andaman district (Fig. 4.9a).

Cliff erosion, sea bed erosion and reef materials from the fringing corals have supplied loose sediments during tsunami waves and storm-induced waves into the shores, which were mostly trapped by beaches of the coves and embayments (Fig. 4.9b). Limestone boulders and larger reef rocks were drifted into the shore

Fig. 4.9 Sediment supplied from the different components of the reef islands into the shore fringes of South Andaman. **a** Sedimentation into the tide pools and rock pools of the shore platform, **b** cliff erosion and mass wasting with limestone dissolution processes, **c** large boulders and sand cay sediments are drifted along the shore fringe platforms during a tsunami event and **d** lithified and cemented rudstone materials along with coralline sands dumped into the landward edge of the shore platform

platforms by the great tsunami event (2004). Sediment filling was also very significant into the subsided estuaries of Katchal Island and Trinket Island of the Nicobar group in a similar event.

From the images and field photographs, it is viewed that billions of forest logs were drifted and trapped by the sediment fillings into the 'Ria' type submerged shores. Many bars and beach ridges were developed by the supply of sand-sized sediments into the mangrove shores of Little Andaman during this period (Fig. 4.9c, d).

The coral fringe coasts of Interview Island, North Sentinel Island, Landfall Island and North Reef Islands of the Bay of Bengal fringe shores were uplifted by 1.0 m or more during the great Andaman earthquake (2004). As a result of such upliftment, the coral reefs of the shallow marine environment were raised above the tide level and therefore added to the island fringes as newly emerged lands. Similarly, many areas of the linear valley flats under intertidal mangrove swamps were also raised above the high tide levels by the seismo-tectonic impacts in the Andaman mainland archipelago, which were also added as newly emerged lands.

Chapter 5
Antecedent Karst Morphology and Future Responses to the Changing Environment

5.1 The Antecedent Karst Morphology of the Reef Islands

The shore fringe barrier reef platforms of a few islands provided substantial favourable conditions in the underwater depressions for coral growth (Fig. 5.22). These limestone platforms are usually visible at low tide and partially covered by marine deposits that dip at 5–7° towards the rim of the islands and extend 60–100 m from the shore cliff. Various reef terraces with morphological features casted on the reef platform surfaces identified as rillen karren, grooves and spurs, lappies, taffoni marks, sharp-edged pools, potholes and seepage pits represent antecedent karst in the shore fringe areas of the islands. The limestone-paved surfaces of shore platforms with antecedent karst morphology are modified by wave abrasion, saltwater weathering and evaporation rates and also partially veneered with single coral blocks, coral rubbles and coralline sand cays at present. They are observed in North Reef Island, Interview Island, North Sentinel Island, Neil Island, Havelock Island, and Trinkat Island of the Andaman and Nicobar groups. As a result of cyclone surges, tsunami surges, swell waves and seismo-tectonic events, sediment was transported along the shores and across the shores of the reef islands to cover up the carbonate surface (Fig. 5.1a–f).

Several outcrops of limestone rocks at Trinkart, More, Kamorta, Katchal, Daanish and Trek Bozo Island of the Nicobar Group along the shorelines represent varieties of fluviokarst, dissolution karst, coastal karst features and etched terrace reef organisms (Table 5.1).

There is a clear signature of a wider limestone pavement surface marked with clints and grikes, joints and various sizes of karren features in the windward side of More and Danish islands under the Little Nicobar group in the Andaman Sea. Expressions of spurs and grooves separated the shorelines into irregular forms, and the inner platform represented a series of doline depressions by the activities of wave abrasions and solution processes in the limestone pavements. After the events of the Great Andaman earthquake (2004) and tsunami waves, the cover-up deposits were

A. Paul et al., *Karstified Coasts of the South Andaman Islands*,
https://doi.org/10.1007/978-3-032-10299-7_5

Fig. 5.1 A few antecedent karts morphology of the Andaman and Nicobar group of Islands: **a** Pinnacle karst at the shore fringes areas of Kalapathar, Little Andaman; **b** grikes and clints on the shore platforms of Little Andaman; **c** Andaman Flysch with grikes and clints and patch reefs, Neil Island; **d** erosional reef terraces of Sitapur, Neil Island; **e** doline depression at the shore fringe area of the limestone platform, Kalapathar, little Andaman; and **f** coastal Karren features of the limestone pavement surface, Buttler Bay, Little Andaman

removed, and the underlying karstified limestone pavement surface was exposed to further wave attacks in the region.

Deeper furrows and narrow ridges over the gently gradient cemented shore platform are visible in emerged and submerged conditions to the southeastern shore of Neil Island (Fig. 5.1). The karstified fossil platform terraces are depleted by wave abrasion in a section of the shoreline that represented the residues of Pleistocene reefs and weathered by acid rains in the region. Some spectacular vertical ridges of pinnacle karst are found in the submerged platform of the southeastern shores, in

Table 5.1 Major antecedent karst morphology of a few islands (Andaman & Nicobar Group of Islands)

Sl. No.	Island sites	Location	Antecedent karst morphology	Remarks
1	Trupthi Island	13° 10′ 3.21″ N 92° 47′ 39.32″ E	Buttressed banks, lapies, raised reef terraces, solution channels, basins, tunnels, etc	Andaman Sea Island, North Andaman District
2	Khurmadera Beach	11° 39′ 55.79″ N 92° 35′ 43.17″ E	Erosional reef terrace, grikes and clints, pinnacle, solution channels, southern basins	Bay of Bengal fringed shores of South Andaman
3	North Sentinel Island	11° 35′ 7.17″ N 92° 12′ 26.58″ E	Drowned basins, solution channels, limestone pavements, erosional reef terraces, etc	Bay of Bengal fringed Islands, South Andaman
4	Neil Island	11° 49′ 44.95″ N 93° 0′ 55.93″ E	Erosional reef terrace, drowned reefs, pinnacle reefs, grikes and clint	Andaman Sea Island, South Andaman
5	Havelock Island	11° 59′ 55.68″ N 92° 55′ 39.35″ E	Limestone pavements, karren, grikes and clint, erosional reefs terraces. solution channels	Andaman Sea Island South Andaman
6	Little Andaman Island	10° 35′ 39.04″ N 92° 32′ 35.63″ E	Erosional reef terraces, limestone pavement, grikes and clints, pinnacle karst	Andaman Sea and Bay of Bengal fringed shores of South Andaman
7	Trinket Island	8° 6′ 58.40″ N 93° 34′ 3.55″ E	Three stepped grikes and clints, limestone pavements, dry valleys, ria type shores, reef flats	Andaman Sea, Nicobar Group
8	Little Nicobar Island	7° 14′ 43.34″ N 93° 38′ 44.88″ E	Elongated shoreface depressions, limestone pavements, coastal karrens, dry valleys	Andaman Sea, Nicobar Group
9	Great Nicobar Island	6° 46′ 8.41″ N 93° 48′ 33.85″ E	Collapsed dolines, limestone pavements, solution channels, dry valleys, grikes and clints	Andaman Sea and Northern Indian Ocean, Nicobar Group
10	Hermit Crab Island	6° 50′ 49.75″ N 93° 47′ 54.02″ E	Limestone pavements with karrens, dolines in collapsed forms, drowned basins, pinnacle karst	Andaman Sea, Nicobar Group

which they emerge usually during the low tide conditions across the gentle sloping platform. There are living corals in the shallow water channels or liner depression between the vertical ridges in the region.

Various types of solution notches, caves, solution pits, doline depressions, grikes and clints, are well preserved in the sections of shell limestones along the shoreline of Havelock Island and Little Andaman. Sea level oscillations, seismo-tectonic subsidence, subaerial weathering, erosion and wave abrasion explain their formation. The conical karsts were possibly developed as antecedent foundations to the reef flats, and the collapsed dolines were developed by the presence of extensive subterranean dissolution zones of the 'blue holes' in the region.

Table 5.1 is prepared in this section of the work to highlight the presence of the antecedent karst morphology in the coral reef fringed shores of the South Andaman group and the Nicobar group of islands in Indian Territory (Fig. 5.1e). Exposures of the island shore platforms to the subaerial environment during the events of low tide or during the lower sea level accelerated karstification processes by the weathering impacts over the limestone pavements of the tropical coasts in the Indian Ocean Islands. The Holocene Sea level rise in the region provided cover deposits of carbonate sediments of clastic materials on the karstified shore fringe platforms, which are reworked and eroded in many places of the archipelago to expose the antecedent karst features on the outcrops of limestone rocks.

The antecedent Pleistocene basement reef limestones are exposed on the stratigraphic sections of the cliff wall at the islands of Ritchie's Archipelago (Neil, Havelock, Ross, Sir William Peel Island, Henry Lawrence Island, Outram Island, etc.) in the South Andaman district, from which the nature of limestones and chemical weathering impacts with karstified features are visible (Fig. 5.1). The boulders of massive limestones from the underwater coral banks were also dislodged and drifted on the shore fringe wave abraded or wave plantation surfaces of the Havelock Island after the events of the great Andaman earthquake (2004) and associated tsunami impacts on the Andaman Sea. These boulder stones and their radiocarbon-dated records reveal that the massive limestone boulders are older (8700 YBP) than the age of the sea-facing cliff failure rock fragments (7500 YBP), and they were drifted from the part of old reef limestone pavements.

The antecedent topographic surfaces recognized in the island are systems of the Andaman and Nicobar groups at the fringes of the Andaman Sea and the Bay of Bengal. They are categorized as: (i) limestone pavements of the older reefs; (ii) eroded reef terraces of the shore platforms; and (iii) island arc margin ophiolite exposures to the basic rocks. The Holocene and the late Holocene reefs are developed over the antecedent topography. Sea level oscillations played a major role in the formation of the Holocene reefs along the Pleistocene older reef platforms in which the signatures of karst morphology are recognized. The karst morphology was developed when the sea level low stands allowed the platforms for prolonged subaerial exposures in the island fringe environment (Fig. 5.2).

The karst forms of the bare limestone pavements are significantly marked by small parallel solution channels, which are usually formed on the seaward inclined rock surfaces under the impacts of subaerial exposures and weathering processes in

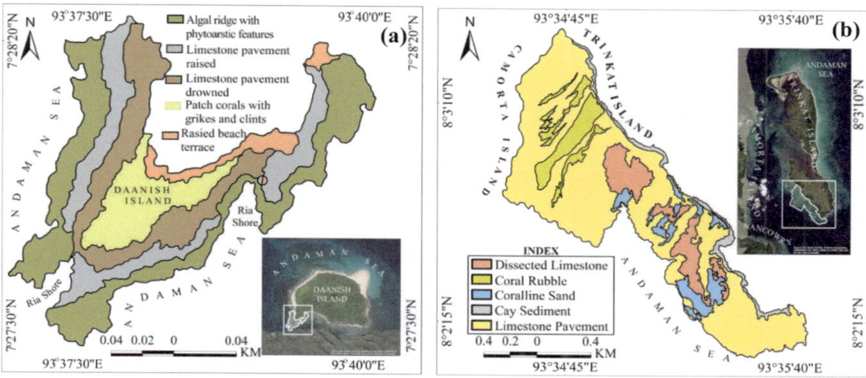

Fig. 5.2 Karstic features identified at the shore platforms with the satellite derived data from **a** Danish Island and **b** Trinkat Island of Nicobar Group

the shore platforms (Figs. 5.2 and 5.3). They are now exposed after the removal of thick cover deposits from the seashores by the concentration of hydrologic energy due to sea level oscillations along the Andaman island fringe environment. Such Rillen Karren features are measured in the field sites of Havelock Island, Neil Island and Little Andaman Island to understand the spatial diversities of the Karst forms (Table 5.2, Figs. 5.4 and 5.5).

Under the influences of precipitation, surface runoff concentration, presence of softer limestones, surface slope in degrees and solution weathering processes, these small parallel features are developed during subaerial exposures in the seashores of the South Andaman Islands (Figs. 5.1a–d, 5.4 and 5.5).

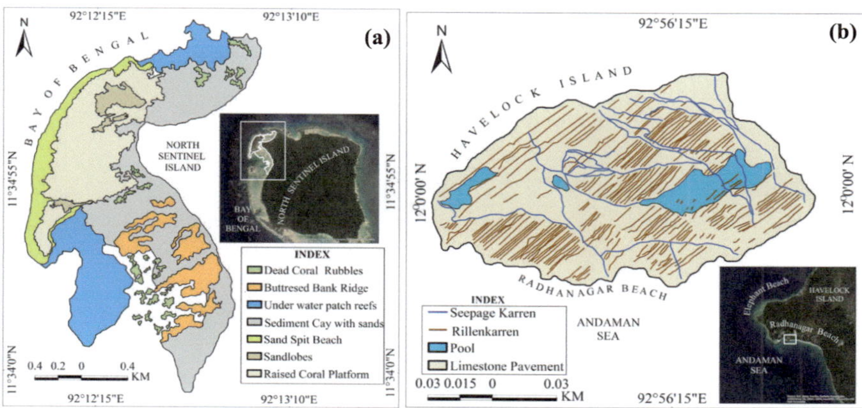

Fig. 5.3 Karstic features identified at the shore platforms with the satellite derived data from **a** North Sentinel Island and **b** Havelock Island of South Andaman district

Table 5.2 Forms of the Rillenkarren features measured in the field survey works at the South Andaman Islands

Location	Lithology	Width (cm)	SD	n	Depth (cm)	SD	n	W/D	Length	SD	n	Mean rill length	Surface slope
Havelock Island	Bare limestone pavement surface	15	1.15	8	12	0.991	8	1.25	500	216.2	8	557.5	7
		18			10			1.8	550				8
		16			11			1.4545	650				7
		15			10			1.5	60				8
		17			10			1.7	600				9
		15			12			1.25	650				12
		16			12			1.3333	700				12
		18			12			1.5	750				12
Neil Island	Calcareous mudstone with algal films	12	1.55	8	18	1.885	8	0.6667	400	76.8	8	528.75	6
		14			19			0.7368	480				6.5
		16			18			0.8889	500				6
		15			18			0.8333	650				6
		15			22			0.6818	600				7
		16			20			0.8	550				7.5
		16			22			0.7273	500				7
		17			22			0.7727	550				7.5
Little Andaman Island	Bare and jointed thick limestone pavement surface	25	1.73	8	10	1.069	8	2.5	1200	33.35	8	1203.75	6
		26			12			2.1667	1250				6.5
		22			10			2.2	1150				6.5
		22			10			2.2	1180				6
		24			12			2	1200				6
		22			10			2.2	1250				6

(continued)

Table 5.2 (continued)

Location	Lithology	Width (cm)	SD	n	Depth (cm)	SD	n	W/D	Length	SD	n	Mean rill length	Surface slope
		24			12			2	1200				7.5
		26			12			2.1667	1200				7.5

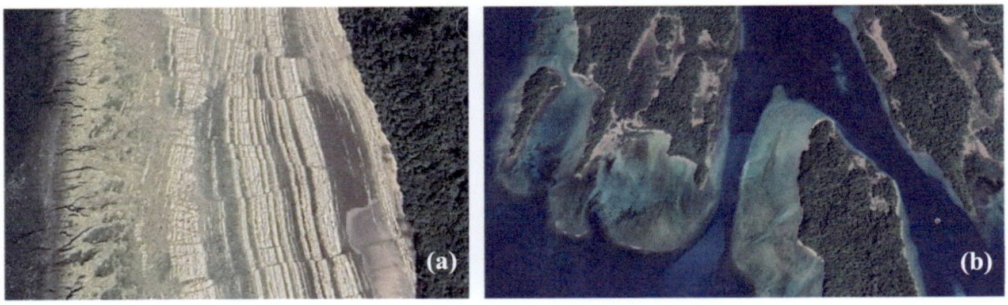

Fig. 5.4 Antecedent karst features of the Andaman Group of islands: **a** Coastal rillenkarren features at the gently shore platforms of Khurmadera beach of South Andaman and **b** the solution channels and basins at the downed reef flats in south Andaman (Jolly Boy and Red Skin Island) district

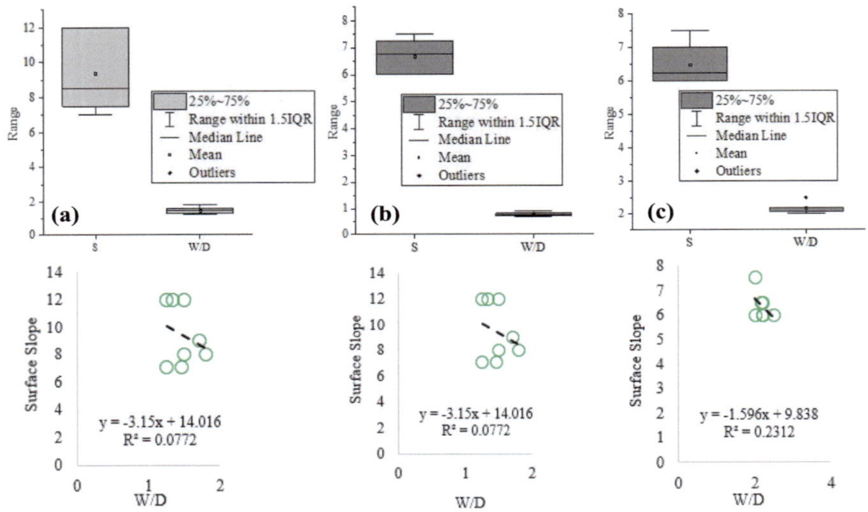

Fig. 5.5 There is a declining trend of liner relationship between surface slope and width-depth (W/D) ratio of the rillen karren features from **a** Havelock Island; **b** Neil Island to **c** Little Andaman Island

The region is now located in the intertidal shore platforms, and the solution channels are wave-abraded and smothered in the forms. In places they are relatively deeper by the frictional process of the pebble and gravel-sized coral rubbles dragged from the low water level to the high water level at the tidal surges, swell waves and storm surge waves.

In Neil Island, at the south-facing windward side, a large tract of the shore platforms is converted into deeper channels that are inclined seaward in a southeast direction and separated by sharp ridges or pinnacle karst features. The framestone, bindstone and rudstone of the jointed limestone are cut into the platform due to solution and wave abrasion processes in the subaerial exposures. The channels are widened and filled by coral rubble but provided attachment surfaces for modern coral reefs and calcareous animals under shallow water levels of the sea.

In Havelock Island, extensive rillen karren features are developed over the jointed limestone pavements of the older reefs, and the karrens are shallower and smother in the intertidal zones. The moisture surfaces of the channel beds are encrusted and binded by green algae mats. They were deeper in forms but later filled up with the coral rubbles, sand rubbles and bio-clastic sands at the tidal zones. However, the channels are more elongated in the limestone pavements of the Little Andaman Island (Buttler Bay), in which the forms of the karst features are sharply modified by oscillation of the sea levels.

5.2 Activities of the Bio-encrustation in the Carbonate Platforms

Among the reef-building framework phases, bio-encrustation plays an important role in reef-building sediments, in calcification rates and in biogenic erosion of the reef rocks. Surface encrustation by thick algal mats supplies algal debris; rock boring animals attached with the reef platforms and other rock bodies produce biogenic erosion; the degraded bodies and fragments of the encrusted animals supply shells and foraminifera into the carbonate sediments as well as other bio-clastic materials in the reef building system (Fig. 5.6).

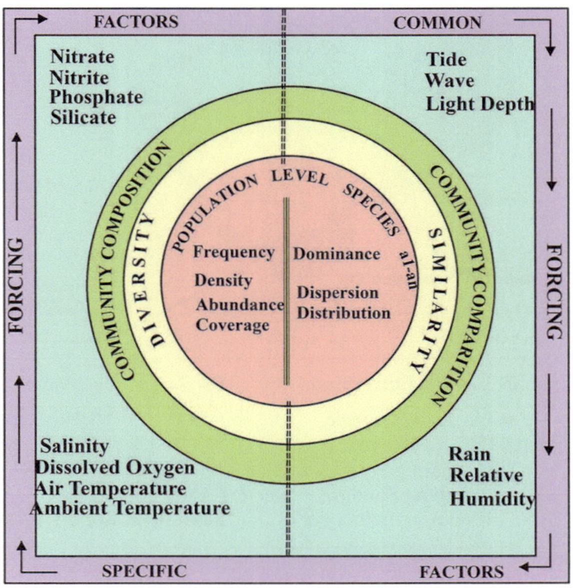

Fig. 5.6 The model showing factors responsible for Habitat dynamics and favourable conditions for bio-encrustations in the Island systems of the marine environment; modified after FDPC, Andaman

Various types of bio-encrusted features are identified and mapped on the section of carbonate platforms in Neil Island. Usually, they have extended in patchy growth and linear growth on the basis of surface Morphometry of the carbonate platforms. However, extensive growth of algal mats occupies most of the carbonated platforms due to continuous moisture supply through the tides and waves. They have been identified as island fringe forest lands, vermetid shell ridges, oyster beds, rock boring signatures on the coralline rubbles, encrusted corals on the rock pools, sea grass patches on the larger pool water bodies, algal mats, lichens on the cliff wall rocks, scarplets with encrusted algaes, etc. in the tropical coast. The high rate of bio erosion is found on the softer limestones and marls along the lower intertidal zones of the island. Decomposition of rocks is possible with bio-chemical weathering processes induced by the activities of such bio encrustation over the carbonate shore platforms. During the grain size analysis of sediments along the cays and beaches, about 21% of bioclasts are identified in most of the samples collected from Neil Island (Figs. 5.6, 5.7 and 5.8).

Number of geochemical properties of sea water is assessed with the water analyser along the near shores of Neil and Havelock Island. Water temperature, water pH, water salinity properties of Havelock, Neil and Little Andaman are compared in this section of the research. At the same time, the soil temperature, air temperature and soil

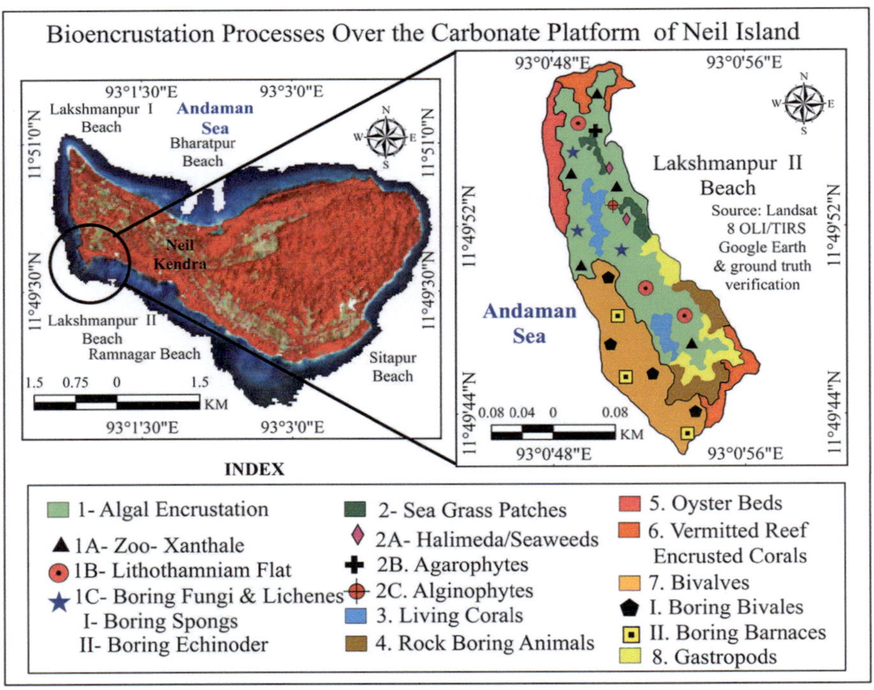

Fig. 5.7 The intertidal biological activity zones over the shore platforms around Neil Island

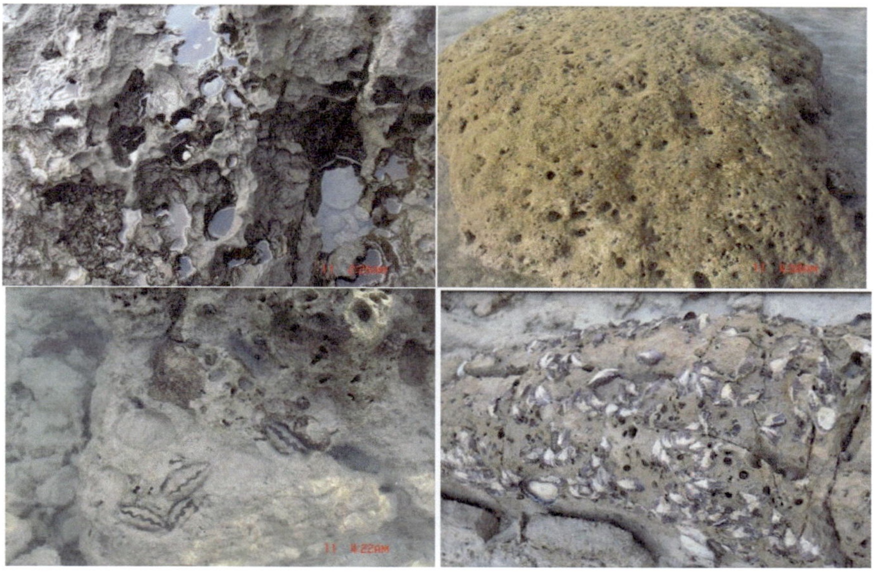

Fig. 5.8 Bio-encrustation on the marls, limestones and coral debris on the shore platforms

pH are also considered for temporal and spatial analysis for all three islands during the field works (20,152,020). The field data of water temperatures for Havelock Island, Neil Island and Little Andaman showing 26.5 to 32.2 °C temperatures in the various sample sites (2015–2020). Such high and abnormal temperatures (February) are reasonably responsible for frequent development of low-pressure centres over the Andaman Sea. Water pH for the same year also represents 8.50 to 9.35 pH over the Andaman Sea around all the islands, which are still alkaline in character. Water salinity for Havelock Island during the period between 2015 and 2020 exhibited significant increase from 24.5 to 26 ppt in the region.

The presence of the moisture, availability of the nutrients, other ecological factors and various attachment surfaces provide ideal conditions for surface encrustation by the blue green algal communities, lichens, various shells, gastropods, mussels, oysters, chitons, barnacles and other rock borers into the carbonate shores of the tropical oceanic islands (Fig. 5.8). The sedimentary matrix of the lithified surface and unlithified sediment cay banks also evidencing the presence of broken fragments of coral rubbles, marine shells, mussels, gastropods and foraminiferas (Fig. 5.8). Large clamps, bivalves and oysters are encrusted within the reef rocks, shell limestones, reef banks at the low tide level of the shore platforms, vermetid shell ridges, vermetid gastropods and the barnacles are encrusted in the bed of tidal pools. Smaller shells or bivalves bore into the rock ribs and old reef stones at the younger stage and died into the rocks after their body growth was enlarged into rock. Which in the later stage produce micro pits after weathering into the rock surfaces in the forms of honeycomb structures (calcareous sandstones, shell limestones, and calcareous mudstones). Over the fragmented limestone pavement rocks, a large number of oval-shaped pits or holes are evidenced by the rock borer 'chitons' in the shore platforms across the tidal zone of Neil Island and Havelock Island. The fore reef banks and the rocky tidal shores are marked by thick oyster beds at the low tide zones all along the shore fringe environment of South Andaman district.

The patches of periwinkles and vermetid gastropods are encrusted over the vertical wall rocks of the shore cliffs, particularly along the seepage points and fissures of the Andaman Islands. Large-size limestone boulders drifted along the shore platforms during tsunami surges (2004) in the islands in which oyster beds, shell ridges and barnacles are attached and encrusted into the tidal zones. Other forms of encrustation are evidenced by the lichens on the cliff wall rocks, logs and fissure cave roofs, blue-green algae on the shore platforms, boulder surfaces, coral rubbles and limestone pavements. The algal ridge on the platform crest is dominated by brown algae and green algae mats, as the encrusted surface is wave-beaten, calcified and lithified carbonate sediments, the features are converted into Lithothamnion flats at the seaward edge of shore platforms in the islands (Figs. 5.9 and 5.10).

Bio-encrustation produces weathering, erosion and accretion processes over the carbonate rocs in response to submersion and emersion of tides in the shore fringe

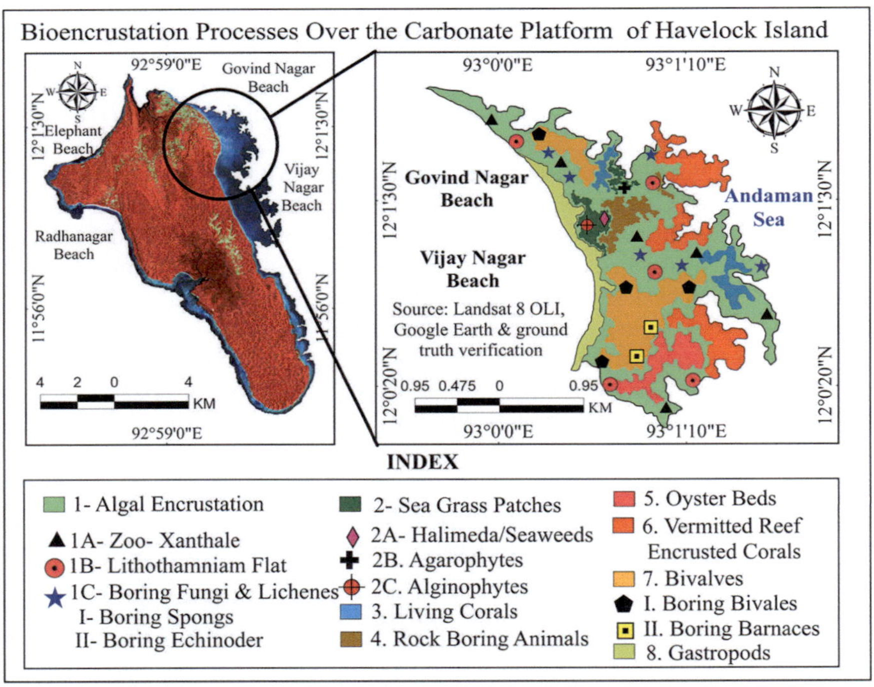

Fig. 5.9 The greater activities of biological processes related with multiple bio-encrustation features in and around the shore platforms of Havelock

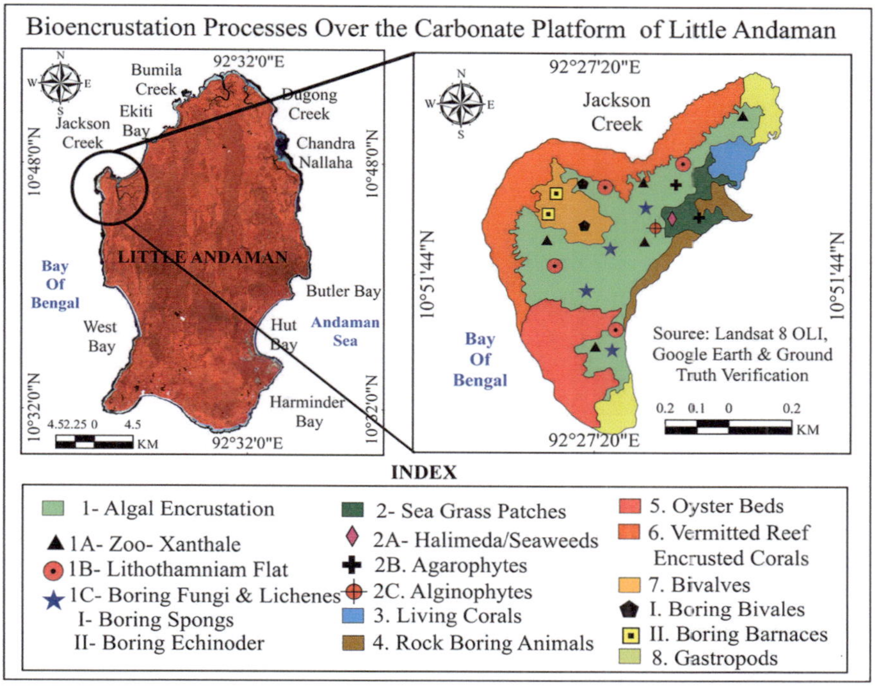

Fig. 5.10 The greater activities of biological processes related with multiple bio-encrustation features in and around the shore platforms of Little Andaman

carbonate platforms (Fig. 5.11a–d). Many residual promontories between the coves and bays of the shore fringe environment are highly affected by bioerosion due to the activities of bio-encrustation as well as by wave abrasion and biochemical degradation. The thick encrusted algal ridges not only resist wave abrasion but also absorb atmospheric carbons and increase the calcification rates within the shore platforms. Many forms of karstic features are promoted by the action of bio-encrustation in the carbonate rocks in the tropical oceanic islands.

Table 5.3 describes types of bio-encrustations, attachment surfaces. Shore platform component, tidal zones, and various characteristics of taphonomy in the shore fringe environments of the archipelago (Andaman and Nicobar groups). There are 18 dominant components of the carbonate shore platforms, which demonstrate the various ecological niches of the shore fringe habitat and are thus occupied by organisms in the marine ecosystem. Island platforms, on the other hand, are occupied by organisms like tropical rainforest vegetation, littoral forests, swamp forests, wetland vegetation and freshwater drainage ecosystems.

Fig. 5.11 Diversified bio-encrustation features at the shore platforms of South Andaman Islands: **a** rock-boring Bivalves; **b** vermetid shell ridge with patch corals; **c** boring pits by the marine shells into the reef rocks; **d** Chitons boring into the reef rock; **e** algal encrusted platforms at the shore fringe environment and **f** oyster ridge attached on the ophiolites exposure in the reef bank of South Andaman Islands

5.3 Environmental Changes and Future Geomorphic Responses

There are three types of antecedent reef platforms in the South Andaman, Little Andaman, Little Nicobar, Trinket Island and Great Nicobar Island (Fig. 5.11e, f). They are explored with field surveys and satellite-derived data analysis in the islands from 2003 to 2018. Holocene reefs are grown in the Pleistocene reef limestones of the island fringes with the oscillations of sea levels and hydrologic controls. The future geomorphic responses of the antecedent reef platforms are discussed in this section with consideration of present-day environmental changes.

(a) Eroding reef terraces at the intertidal shore platforms with karst features;
(b) Submerged reefs with karst features in the nearshores;
(c) Karst features below the cover deposits of the shore platform limestone pavements along the island margins.
(d) The reef terraces attached with the island fringe lands contain fossil reefs, solution cracks, solution channels, solution pans, and steep sided slopes, which are affected by the wave abrasions and water-layered weathering process (Fig. 5.22).

When the sea level was relatively higher in the region, the island-fringed reef grew under favourable conditions, but a slight fall of the sea level brought it under subaerial exposures to develop karstic features by the impact of precipitation and the active solution process. The dead reefs were further raised by the past events of seismo-tectonics and by the thrust belt tectonics in the island fringe environment. However, the present-day rising sea level with frequent events of surge waves on the

Table 5.3 Types of bio-encrustations in the shore fringe carbonate platforms of the South Andaman Islands (LTL—low tide level; HTL—high tide level, MTL—mid tide level, STL—supra tidal level)

Sl. No.	Shore platform zones and dominate components	Tidal zones	Associated biota	Characteristics taphonomy
1	Fore reef banks	Nearby LTL	Bivalves, clams, oysters barnacles	Solution channels, pits and basins
2	Drifted boulders	Nearby LTL	Barnacles, bivalves	Micro-pits
3	Rims of the shore platforms	Nearby LTL	Brown algae, oysters	Microbial crust
4	Limestone ridges in promontory	Nearby LTL	Rock boaring shells, scallops	Solution pits and fissures
5	Solution channels	Nearby LTL	Blue-green algae	Lithified bioturbated holes
6	Coral rubble ridges	Between MTL	Algae, boring shells, mussels	Micro-pits
7	Solution basins	Between MTL	Vermetid gastropods, lichen	Microbialites crust
8	Erosional reef terrace benches	Between MTL	Sea grasses, patch corals	Micro-pits, bio-erosion
9	Tide pools	Between MTL	Algae beds, clams	Microbialites crust
10	Shore platform surfaces	Between MTL	Foraminifers, molluscs fragments, bio-clastic sands	Cemented spikes, solution pits
11	Beach sands	Nearby HTL	Foraminifera, hermit crabs gastropods	Bio-accretion
12	Beach rocks	Nearby HTL	Algae beds, foraminifera	Bio-erosion
13	Sediment cay banks	Nearby STL	Foraminifera, molluscs fragments, bio-clastic sands	Bio-accretion
14	Shore cliffs walls	Nearby HTL	Lichens, vermetid gastropods periwincles	Wave splashing marks tafoni marks solution
15	Notches	Nearby HTL	Gastropods, periwincles	Solution pits
16	Sea caves	Nearby HTL	Lichens, molluscs shells, brachia	Solution pits
17	Fragmented pavements	Nearby HTL	Chitons, algae mats, shells	Bioerosion
18	Stacks, pinnacle karst	Nearby HTL	Barnacles, lichens	Abraded skeletal features, solution pits and fissures

Fig. 5.12 Karstified reef terrace eroded and fragmented into the shore fringe areas of Lakshmanpur, Neil Island

windward sides produces energy concentration to erode the reef terraces slowly and steadily in the region (Fig. 5.12).

The steep-sided erosive walls of the reef terraces have led to knowing the exposure sites for identification of ancient reefs with karst features in the several islands of South Andaman and Little Andaman. With time and a changing environment, the wave-abraded residues of the reef materials will be dumped into the high tide line, and gradually the rounded pebble beach and shingle beach of cover deposits may be developed over the karstified limestone pavements in the intertidal shore platforms. Remnants of the reef terraces are already separated into isolated ridges or stone pinnacles by water-layered weathering and wave abrasion and solution processes by precipitation. In the future, with sea level oscillations, the entire reef terraces will be cleared by erosion and weathering processes.

(b) There are several drowned reefs in Neil Island, Havelock Island, Little Andaman Island, Trinket Island, Little Nicobar Island, Katchal Island and the parts of the Great Nicobar Island in which several karstic features are identified. The karst features occurred when the sea level lowstand allowed the island platforms for prolonged subaerial exposures in the region of the Andaman Sea and Bay of Bengal fringe environment.

Features like solution channels, solution basins, jointed limestone pavements, solution rims, furrows, spurs, and grooves, residual clints, ridges or pinnacle karsts, and wide grikes are found in the shallow water reef banks at present. With the changing environmental conditions of the hydrological processes and the effect of seismo-tectonic sediment supply, the karst features of the antecedent platforms will be filled up with a cover of sediment deposits. The colonies of modern reefs will spread in patches with available favourable conditions in the drowned reef platforms. Such colonies of patch reefs are explored in the former karst basins of the drowned reef platforms in the island fringes of South Andaman.

Fig. 5.13 Eroded reef terrace of Ramnagar shore platforms, Neil Island

During the events of swell waves, surge waves and tsunami waves, the shallow water drowned reefs were affected by sediment deposition and removal of zooxanthellae in the region. The oscillations of sea level will decide the fate of drowned reefs with the capacity of catch-up, give up and keep-up methods of the reefs (Fig. 5.13).

(c) The older reef limestone pavement surface under the blanket of cover deposits with supply of coral clasts showing karst features in the intertidal shore platforms where the veneer deposits were removed by the sea waves in the surge levels (Fig. 5.13). The bare limestone pavements are jointed, fractured, karstified, and characterized by developed grikes and clints, Solutions channels, as well as taffoni ferrous in the shore platforms of the island fringe environment. Several phytokarstic features like spikes, solution basins, algal ridges, taffoni marks, coastal karren, solution pits, grooves and primary lappies with bio-encrusted surfaces are found in the intertidal shore platforms.

The karstified limestone pavement surface was exposed above the sea level when the subaerial weathering process activated by the precipitation and seepage waters from the island surface produced such phytokarstic features with time and the minor oscillations of the sea level that the limestone pavements came under the intertidal environment and gradually fielded up with cover deposits by the supply of carbonate sediments and coral sediments from the fore reef banks in the island fringe areas. The colonies of patch reefs and sea grasses are found to develop into the rock basins, intertidal rock pools and other watery patches of the shore platforms.

The patch reefs of the pools of water bodies are adversely affected due to sediment deposition and boulder drifts across the shore platforms in the great Andaman earthquake and tsunami events (2004). Such older reefs of the limestone pavements will be trapped again by the cover deposits of organic, carbonate and siliciclastic sediments in the near future with rising sea level trends.

(d) Different sizes of fragmented limestone pavements and dislodged limestone boulders with karstified features are distributed in the intertidal shore platforms from the fore reef banks after the events of dynamic hydrologic processes influenced by various energy drivers (wind waves, cyclone surges, tsunami waves, ENSO regimes, etc.) in the oceanic islands. Such dislodged boulders and limestone pavements showing the signatures of karstic features like solution pits, channels, solution basins, lappies, bio-encrusted surface, tafoni marks in their bodies.

They also represent the features of antecedent reefs in the shallow water island fringe environment of Andaman group of islands. With changing environment, the fragmented older reefs will be blanketed by cover deposits of coral sands, carbonate sediments and drifted coralline debris in the shore fringe areas. They may be converted into raised beach features or reef terraces by seismo-tectonics and thrust belt tectonics in the region.

5.4 Karst Features as Geotourism Sites in Andaman

Karst islands of Andaman are notably used as attractive geotourism sites for their geological, geomorphological and geochemical significances in the tropical moist climate. The geotourism sites of important karst features are identified as: limestone ridges at the headlands, and promontories, limestone caves, tower karst, karst doline basins, pinnacle karst, limestone cliffs, eroding reef terraces, karst tunnels, karstified beach rocks, limestone pavements with karst features, marshy or swampy forests over the limestone platforms, karstic blowholes in the shore platforms, cay sediment bank or reef of sands, sink holes, etc. in the Andaman and Nicobar group islands. Such natural architectures of the limestone terrains can be categorized on the basis of their value assessment (aesthetic values, tourism values, geo-diversity values, ecological values) for sustainable tourism, conservation efforts and community engagement in minimizing the environmental impacts (Table 5.4).

During the perception survey about the significances of the natural geo-architectures in the wild landscapes, the tourists from other countries replied, 'only the lucky persons can find the features build by nature through architectures on the landscapes over the period of time'. They feel the beauty and quality of natural architect of the landscapes, and favour restoration in the natural heritage features in such oceanic Islands. Tourism operators have considered only the limestone caves and caverns, natural arches or bridges, and cay sediment banks in a few islands to attract tourists in the geotourism destinations (Figs. 5.14, 5.15, 5.16 and 5.17).

Both the tourists and tour operators have very little ideas or understandings about the karst features. Visitors or tourists and the tour guides only know the limestone caves of Baratang Island, and Howrah Bridge (Natural Arch) of the Neil Island. The nature tourists and excursion tourists only understand the significances of the karst features in the tropical oceanic islands. Growing mass tourists will disturb the conservation and restoration of such natural heritage features in the allowed destinations. The present study will exhibit the specialized karst features with value addition

Table 5.4 Geotourism significance of karst features in Andaman and Nicobar Islands

Destination sites	Geomorphological value addition (V_{GEO})				Geomorphic value (V_{GEO})	Management approaches
	Aesthetic landscape value (Est.)	Cultural value (cult.)	Scientific value (Sec.)	Ecological value (Eco.)		
Little Andaman, Havelock Island	3	6	5	2	1.8875	Environmental Management
Baratang Island, Neil Island	2	8	5	3	2.375	Heritage Management
Little Andaman, Moore island	2	5	6	2	1.635	Heritage Management
Little Andaman, Trinket Island	3	4	4	2	1.5775	Tourism Management
Little Andaman, Neil Island	2	2	3	3	1.455	Heritage Management
Havelock Island, Neil Island, Munda Pahad	3	8	5	2	2.1875	Tourism Management
Neil Island, Khurmadera Beach	2	6	4	2	1.765	Environmental Management
Little Andaman, Baratang Island	4	7	7	1	1.87	Heritage Management
Neil Island, Chidiyatapu, Wandoor	2	4	5	1	1.175	Tourism Management
Buttler Bay, Havelock Island	3	6	4	2	1.8775	Environmental Management
Ramnagar Beach, Vijaynagar Beach	7	7	5	5	3.3875	Environmental Management

(continued)

Table 5.4 (continued)

Destination sites	Geomorphological value addition (V_{GEO})				Geomorphic value (V_{GEO})	Management approaches
	Aesthetic landscape value (Est.)	Cultural value (cult.)	Scientific value (Sec.)	Ecological value (Eco.)		
Neil Island	6	7	4	4	2.965	Tourism Management
Little Nicobar Meroe Island	2	5	4	2	1.615	Heritage Management

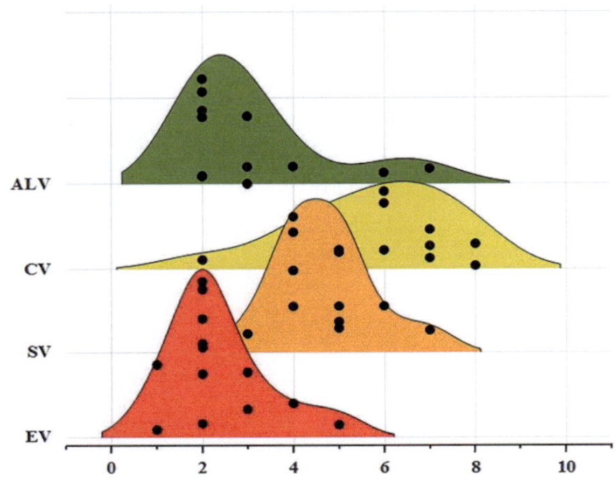

Fig. 5.14 Destination sites wise value addition for the Geo-tourism significance of the Karst features in Andaman and Nicobar Islands (AL-Aesthetic Landscape Value (Est); CV-Cultural Value (Cult); SV-Scientific value (Sec) and EV-Ecological value (Eco))

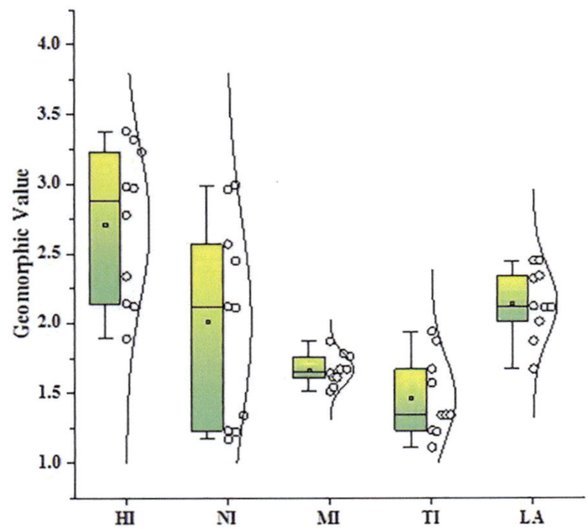

Fig. 5.15 Geo-morphosites index values for the five different islands (HI—Havelock Island; NI—Neil Island; MI-Moore Island; TI—Trinket Island and LA—Little Andaman) of Andaman and Nicobar groups estimated from the data presented in the Table 5.4

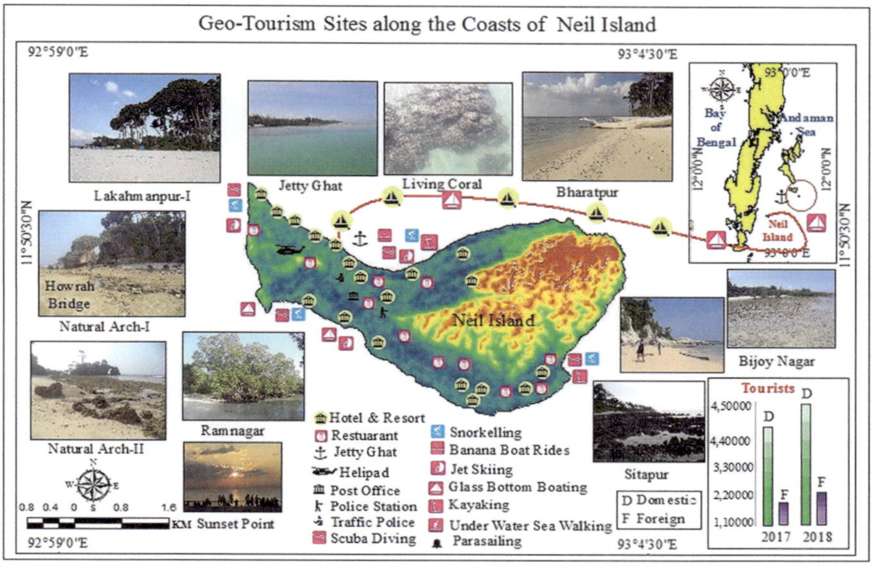

Fig. 5.16 The multiple sites of geo-tourism areas played as the magnet of attraction in Neil Islands

components for restoration, conservation and management of the geo-tourism sites in the islands (Figs. 5.16 and 5.17).

There are some management approaches to restore the natural heritage sites of karst features in the island environment in a sustainable conservation. Following recommendations are proposed and formulated to show their importance in geo-tourism destination network of the region:

1. Signage boards with scientific description of the specific karst features should be displayed in the destination sites for learning by the visitor tourists;
2. The tourism department can organize the workshops or training courses for the tour agents, tour operators, and tour guides for greater understanding of the regional karst landscapes and their significances;
3. For sustainable heritage management of the natural architectures one interpretation centre must be there in geo-tourism destination sites of the islands for mentioning the geological history of the development of karst features and indication of possible modifications with the changing environment;
4. A certain amount of the profit earned from the visitors should be utilized for the maintenance of the heritage features;
5. Media exposure of the heritage karst features are immediately needed for attracting the geo-tourists and eco-tourists in the region;
6. There must be the heritage protection guards or care takers of the natural landscape architectures to save them from the harmful impacts of mass tourists.

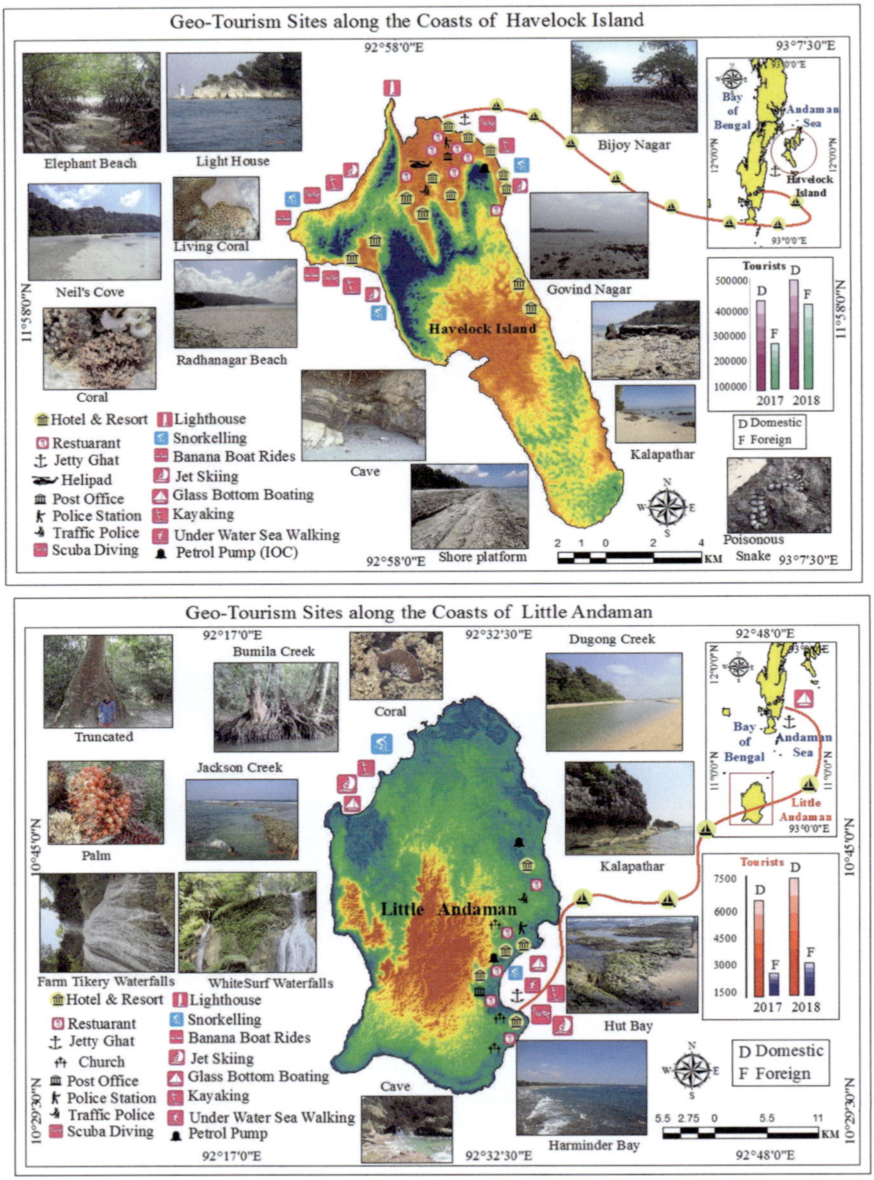

Fig. 5.17 The multiple sites of geo-tourism areas played as the magnet of attraction in Havelock Island & Little Andaman

5.5 Future Climate Change Impacts on the Littoral Vegetations and Coral Reefs

The Andaman and Nicobar group of islands constitute diversities of natural vegetations and forest belts extending from the land–sea interface to the hill ridges and inner parts of the islands. Among them, the littoral forests and swamp forests exist in the inter-tidal and supra tidal shore fringes of the oceanic islands. Mangroves, mangrove associates and other halophytic vegetations are extended in the tidal zones of estuaries, creeks, straits, bays, river mouths and shore platforms and along the sand reefs, sediment cay banks and shore fringe reef terraces of the region of Andaman Sea and Bay of Bengal coasts (Fig. 5.18). The shore platforms and estuary mouths were fringed with dense luxuriant frontal mangroves of carbonate settings and tidal roots (breathing roots, prop roots and cable roots) with wide canopy covers and lianas. Back mangroves are located in the creek banks, channel straits, bay fringe sedimentary basins, carbonate sediment deposited limestone pavements and along the 'Ria Shores', valley flats of the inner islands. Mangroves are also thriving in the landward parts of the raised beach terraces of the supratidal zones.

However, the three belts of zoned mangroves of the differential physiographic settings are more or less affected by the events of earthquakes, tsunami surges, cyclone surges and swell waves in the region. The frontal mangroves of the shore platforms are degraded in Neil Island, Havelock Island, Chidia Tapu, Mundapahar,

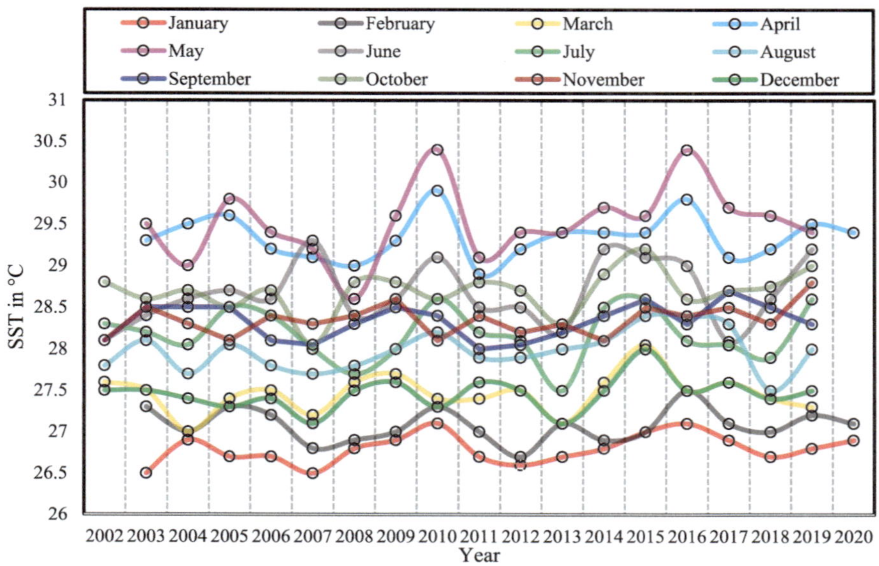

Fig. 5.18 Showing the month wise SST fluctuations of the time span between 2002 and 2020. The satellite derived data set for Andaman and Nicobar Island are plotted to show the variations in SST at the temporal scale

Gorasharma, Chouldharia, Viper Island, Buttlar Bay, Hut Bay, Harmindar Bay, Trinet Island, Katchal Island and Nicobar Island. Allochthonus sediment supply with sand drifts, boulder drifts and coralline debris drifts had produced a thick covered deposits in which the tidal roots of mangroves were trapped in the shore platforms, embayments and river mouths of the crenulated shores.

Back mangroves of the little Andaman Island were badly damaged at the backwaters of raised and blocked river mouths after the Great Andaman earthquake (2004) and tsunami surges. Intertidal mangroves belt have been raised by tectonic upliftment and converted into supratidal mangrove belt above the tide levels along the western banks of Andaman Islands. The Intermontane valley flats mangroves of the 'Ria Shores' along the north–south aligned estuaries were raised above the tide levels and partially killed in the zones of thrust belt tectonics during the events of severe earthquakes. Areas of the back mangroves were also damaged in parts of Baratang Island in the middle Andaman, along with Rangat bay, long island and Interview Island. The cyclonic winds and surge waves generated during the el-Nino and La-Nina phases in the eastern Bay of Bengal and Andaman Sea also produced the impacts over the back mangroves of Andaman Islands.

The littoral vegetations of the shore fringe above tide terraces, sand banks and shore cliffs represent the varieties of mangrove associates and monsoon mixed forests in the oceanic islands. They are very common behind the frontal mangroves and tolerate the salinity of soils in the shore fringe environment. Among them wild jamun, sea mahul, wild cashewnut, pandanus, sea hibiscus, sea poison tress, wax jambo, sea oak tree, India rosewood, Indian beech, Indian mulberry, Padouk trees are common in the South Andaman, Middle Andaman and North Andaman groups of islands. Along with these evergreen trees varieties of creepers, vines, herbs, shrubs, grasses, heaths also found in the sandy banks of the islands (Fig. 5.19).

Usually, the log drifts are very common along the shore cliffs, shore platforms, reef terraces and creek mouths in the island shores by the impact of surge waves, wind storms, cliff retreats, seismo-tectonics and tsunami waves. The subsided valley

Fig. 5.19 Changing wave regimes and active denudational processes on the shore platforms of the Island fringe environment in the windward side of the South Andaman Island

bottom littoral vegetations are heavily degraded in the Katchal Island and Trinket Island of the Nicobar group during the events of earthquake and tsunami surges. The global warming related phenomena, increased rates of SST in the Andaman Sea, frequency and intensity of cyclonic storms and the sea level rise threats will disturb the shore fringe littoral vegetations in Andaman Island by uprooted trees from the cliff tops and drifted long along the shore platforms.

5.6 Carbon Balance and Carbon Stocks in the Phytokarst Landforms

About 40 percent of the global atmospheric carbon is sinking into the ocean fringe coastal wetlands, and over 17% of the atmospheric carbon is also sinking into the landed biosphere to keep the carbon balance other that the oceanic sink (43 percent) of carbon cycle in the earth. Most of the sinking carbon in the form of dissolved carbon dioxide is building the calcareous sediments into the shallow seas and coastal wetlands. However, the uneven karstic landforms of the intertidal shore platforms with algal encrusted surfaces, tide pool sea grass patches, patch corals of the watery basins and irregular frontal mangrove covers by calcification rates, photosynthesis process affected the ecosystem carbon uptake through biomass production and carbon release via decomposition. Thus, at the local and regional scales, the areas of Phyto karstic landforms and biomass features respond to the climate change by feedback system to the atmospheric carbon balance (Table 5.5). The limestone karst terrains on the other hand are also subjected to solution by weak carbonic acids in production of sediments once more in the shore platforms caused by rainwater charged with carbon dioxide from the atmosphere.

The present section represents total amount of carbon stocks along shore platforms of Neil Island using the Remote Sensing data and field measurements data, as well as with the review of existing literatures (Table 5.5, Fig. 5.20). Shore platforms of the South Andaman Island are categorized into six major biomass zones within the areas of tide levels and surge level exposures. As the biomass areas act as carbon sinks, the carbon stocks are assessed separately in the six zones and finally added as the total stock of carbon in the narrow shore platforms of the island fringe environment. Thus the shore platforms of Neil Island plays very significant role in the regional carbon balance by absorbing significant amount of atmospheric carbon. All the six phytokarst regions of the shore platform wetlands need protection and management from the impacts of tourism pressures at present. It is depicted from the study that the patch corals, mangroves, coral beach sands and limestone pavements with algal encrustation a sea grass colonies absorb 16,386,300 mg, 11,735.1 mg, 103,888 mg, 190,735 mg and 739.3 mg carbon per year sequesters respectively in the shore fringe wetlands of the carbonate coast.

Table 5.5 The assessment of carbon absorption rates by the shore platform biomass components (Estimated around the shore fringe carbonate platforms of Neil Island)

Sl. No.	Shore platform biomass component	Area (Sq.km)	Average rates of carbon sinks (mg/m2/yr	Total amount of carbon sinks per year in mg
1	Patches reefs in the smaller water pockets	364,140	45 (Raj et al. 2022)	16,386,300
2	Seagrass colonies in the intertidal rock pools	86,975	0.0085 (Rajamohanan Pillai Ranith, 2024)	739.2875
3	Thick algal encrusted surface in the carbonate platforms	237,521	0.0410 (P. Kaladharan, S.Veena and E. Vivekanandan 2009)	9738.361
4	Mangroves cover areas with carbonate sediments	237,312	0.04945 (Kiran Shah, Mohan P.M. 2020)	11,735.0784
5	Bio-clastic beach sediments with calcareous materials and coral fragments dominated by island	6493	16 (Kiran Shah, Mohan P.M. 2020)	103,888
6	Limestone Pavements with coral rubbles and beach rocks with algal films	6,357,835	0.03 (P. Kaladharan and Reeta Jaya Shanker, 2003)	190,735.05

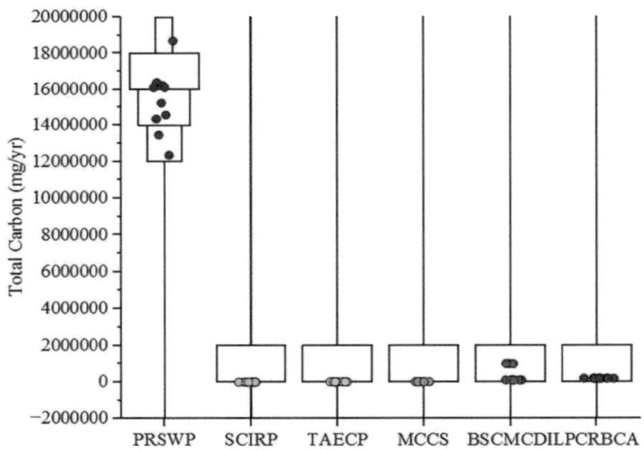

Fig. 5.20 The Box plot diagram showing the shore platform biomass component wise total carbon stock at Neil Island of South Andaman district (PRSWP-Patches Reefs in the smaller water pockets; SCIRP-Seagrass colonies in the Intertidal Rock Pools; TAECP-Thick Algal encrusted surface in the carbonate platforms; MCCS-Mangroves cover areas with carbonate sediments; BSCMCDIL-Bio-clastic beach sediments with calcareous materials and coral fragments dominated by island; PCRBCA-Limestone Pavements with coral rubbles and beach rocks with algal films)

5.7 Ecosystem Functioning and Human Wellbeing in the Island Platforms

The ecosystem components include (i) topography, (ii) limestone, (iii) terrains, (iv) surface drainage conditions, (iv) underground water flows, (v) forest biomass, (vi) agricultural activities, (vii) human settlements, (viii) shore platforms, (ix) fringing coral reefs and the climate variability in the archipelago with maritime influences. They are again categorised into abiotic, biotic and anthropogenic components for explaining their functions and carrying capacities. The modern human systems and the aboriginal tribes are surviving in the few islands on the basis of available resource potentialities in the isolated landmasses under the tropical maritime environment (Fig. 5.21).

To run the ecosystem functions and human wellbeing the resource management, agricultural management, waste management, tourism management and inter island transport communication management are important activities of the local administration. Thus, the carrying capacity of the human modified islands are always assessed by the agricultural department to find out the available resources and favourable space to support the increasing human population and livestock population for sustainable environment and human well-being.

The present section discuss the functions of the forest ecosystem, freshwater ecosystem and shore fringe wetland ecosystem: The island platforms with forest lands contribute soil organic matters, soil nutrients, surface stabilization process

Fig. 5.21 The aboriginal tribes (Onge and Nicobari) living in the littoral forest habitats of the Little Andaman Island after the tsunami event of 2004: **a** an Onge in the Hut Bay shore fringe area; **b** a Nicobari family with their ornaments and utensils; **c** interview survey with the Nicobari people; **d** interacting with a Nicobari in his house; **e** a Nicobari hut and **f** during house hold survey in a Nicobari village of Little Andaman

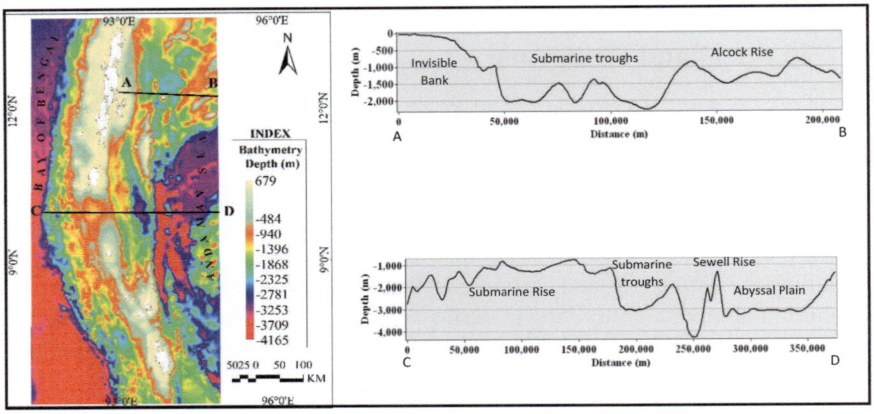

Fig. 5.22 The regional bathymetric data is represented through the cross sections in the diagram to find out the submarine benches across the shallow marine environment and deep marine environment

by their rooting zones over the porous limestones and hold the soil moistures in different land units to support the natural ecosystems in the oceanic environment. The evergreen trees and rain forest trees play most important role in the local hydrological cycle by attracting rain bearing clouds through convective method by increasing the evapotranspiration rates from the dense forests.

The forest biomass of the island platforms absorbing atmospheric carbon dioxides and charging with the rain waters and after mixing with the decomposed organic matters and after mixing with the decomposed organic matters and after mixing with the decomposed organic matters reaching into the limestone soil across the forest ecosystem in the forms of acidic liquid bicarbonates. These elements are more powerful chemical compounds to react with the limestones of soluble materials, and as a result of increasing weathering rates the soluble carbonates are washed into the shore platforms as another source of sediments. The degassing of the bicarbonates at the seepage zone above HTL during low tide condition supply micrites or cements to the bioclastic sands and sand rubbles of the high tide zones to develop beach rocks or lithified sand reefs in the shore platforms. The atmospheric carbons are also mixing with the carbonate crust through the forest ecosystem when carbons are released into the soil by decomposing of trees in the island platforms.

Presence of the forest ecosystem also contributes a role in climate stabilization by absorbing atmospheric carbon dioxide, reducing temperatures and generating the rains by releasing moistures for saturation of clouds above the forest lands. The littoral forests and swamp forests act as buffers against the impacts of waves, erosion, wind storms and tsunami surges in the island ecosystem.

The habitat of aboriginal tribal communities are also well preserved and well protected by the natural forest ecosystem in Andaman and Nicobar groups of Islands. Forest dwellers of the region (Shompen, Nicobories, Jarwas, Ongees, Sentinelese, Great Andamanese) maintained their livelihood by hunting and gathering economy in a primitive methods under the rich biodiversity of the forest ecosystems. Harmful activities played by the modern human society reduced the extent of natural habitats for the aboriginal tribes, thus some restricted habitats are not allowed to visit by the tourists and settlers in the groups of Islands.

Various karst features are well developed in the limestone cliffs, promontories, hill slopes, along the Intermontane valley flats, and near the thrusted hills of the island platforms. The jointed limestones of many islands are affected by erosion and weathering processes which also supply the carbonate sediments into the shore platforms with the seepage waters and runoff waters during rains.

The shore fringe wetlands of brackish water ecosystem, and the drainage fringe valley flats of fresh water wetland ecosystems act as filters of pollutants, waste and sediments in the island platforms. They accumulate organic matters in their beds with sediments and hold moistures along the bank margins to support the other components of the habitats. In the porous limestone terrains fluviokarst features (dry valley, solution notches, solution spikes, dolines), solution basins, wet valleys, wide and open basins are developed. Artificial drainage channels constructed along the hill slopes offer water resources for different uses and recharging aquifers, but usually dry outs in the non-monsoon months.

The annual ground water yield is sensitive and insufficient to support the human populations, livestocks, and agricultural activities at the optimum levels in the islands. As a result of the discharges of rainwaters through drainage runoff, surface runoff and seepage flows across the relief zones into the seashores, the scarcity of freshwater is a major problem of the island ecosystem. Thus, water resources management are considered as priority to assess the carrying capacity of the islands. A large influx of tourists and visitors, and the municipalities demand freshwater supply in the island at present.

Summary

Following findings and conclusions are summarized below on the basis of present study in the tropical oceanic island of northern Indian Ocean around the Andaman Sea and the Bay of Bengal fringed coasts of the Archipelago. The book volume represented the following summaries placed as per the chapter wise findings of the present study.

South Andaman Islands (Physical Settings)

Chapter 1 represents the physiographic settings of the South Andaman district. The Andaman group of islands includes a total of 575 islands distributed between the Bay of Bengal and the Andaman Sea in the northern Indian Ocean. They are separated into 3 administrative districts under North Andaman, Middle Andaman and South Andaman, extending from $13^{0}15'$ North Latitude to 10^{0} North Latitude in the maritime region of the tropics.

The reef limestones with calcareous sandstones, mudstones, shell limestones, white claystones, chalky limestones and marls of the Archipelago group of geological formations and the Pleistocene limestones of the island interior sections are the major sites of karstic landforms found in the South Andaman district. The tropical maritime climate, dense evergreen forests, high rainfall (<3000 mm/yr), intensely jointed and thicker limestones at the active tectonic zones and bio-encrustations of the shore platform components provided ideal conditions for the active karstification processes in the region.

The island interior parts and the coastal fringe areas are explored in South Andaman for assessing the karst features, speleogenesis, limestone dissolution processes and types of karstic landforms in the study. The chapter also demonstrated how far the pseudokarstic features were affected by various marine processes and seismo-tectonics in the Andaman group of islands.

A. Paul et al., *Karstified Coasts of the South Andaman Islands*,
https://doi.org/10.1007/978-3-032-10299-7

The coastal fronts of the islands were largely affected by the events of the Great Andaman Earthquake and the tsunami waves (2004) and cyclonic landfalls (2014) with storm surge waves. During the events of 2004, the reef banks, the reef flats, the shore platforms and the shore cliffs of the limestone rocks were affected by fractures, fissures, faulting and collapsing processes under the subduction zone tectonics and thrust belt tectonics. The vertical displacement of the shoreline was also recorded in the northern and southern islands of the Andaman group.

Coastal Geomorphology and Habitats

Chapter 2 demonstrated the geomorphology and habitats of the South Andaman Island groups. Various land classes and their genesis are studied with geospatial techniques and through the monitoring processes in the rigorous fieldwork conducted during the period between 2012 and 2018, particularly around the rock outcrops of the islands. The shoreline configuration types of the embayed coasts and their stages of development are studied in this section using the satellite-derived data, spatial modelling techniques and monitoring the shoreline geomorphology to identify the role of maritime processes in the evolution of the carbonate coasts.

The fringing coral reef habitats, jointed limestone terrains, shore platform-based mangrove habitats, littoral vegetations, algal encrustations and other bio-encrustations with weathering phenomena of the shore platform components are considered in the present study to represent their role in the formation of phytokarstic features. Reef sediment supply into the shore platforms and their morphodynamics under the island surface and marine processes on the windward and leeward sides of the region are also represented with their local surface morphometry in this section of the work.

The geological and lithological components of the group of islands are explored at the sites of cliff outcrops and shore platform outcrops. Similarly, the nature of carbonate rocks and reef limestones, with their physical components is studied to represent the mechanisms of karst feature development under subaerial weathering and water-layered weathering processes in this part of the study.

The surface morphometry of the shore platforms characterized by topographic depressions, rock pools, tide pools, smaller lagoons and sloping platforms provided the accommodating space for absorbing the supplied sediments from the reef bank, reef flats, shore cliffs and island platforms to run the shore platform morphodynamics with interacting and interplaying processes between physical factors and biological factors of the island environment in the tropical oceanic coasts.

Karstified Coasts and Coastal Hydraulics

Chapter 3 XE highlighted the nature of karstified coasts and the related coastal hydraulics of the oceanic islands. Limestone dissolution, tidal pumping, seepage lines at the upper limit of HTL, water layer weathering, evaporation, precipitation, calcification rates of the shore fringe biomass, supply of calcium bicarbonate acids from the island platforms of wooded landscapes, and bio-encrustation processes are explained, as well as their spatio-temporal variations are demonstrated in this section of the study. A schematic classification of karstified coasts is documented for the South Andaman Islands in this section.

Coastal karst features are identified in the shore cliffs and shorefront platforms of the islands, and their dimensions are measured during the repeated field survey events. Sea caves and island interior caves are explored while monitoring the geomorphological features attempt was made to explain the role of intermontane valley flats and vadose water layers, and oscillating sea levels as base-level control factors in the evolution of caves in the islands. Extensive beach rock formation sites with beach calcarenites and eolianites, calcareous sandstones, shell limestones and conglomeratic debris limestones are found in the high tide zone and mid tide zone of the islands. The island-front seepage zones with the supply of dissolved carbonic acids played an important role in the lithification and cementation of beach deposits at higher sea levels.

Regional sea levels, interannual variations of the water levels, tide levels, storm surge levels, swell waves, tsunami waves and regular wave activities, as well as the discharge of spring waters from the shore cliffs vertical walls during rainy months are considered as the components of coastal hydraulics for playing the role and the promotion of the karstification processes on the shore platforms of the island fringe environment. Phytokarstic features are more prominent in the algae-encrusted biofilm surfaces on the limestone beds of the island coasts.

Reef Islands and Carbonate Sediments

Chapter 4 described the nature of reef islands and the products of carbonate sediments derived from the fringing coral reefs and the supply of weathered materials from the island platforms with the product of mass wasting materials from the shore cliffs to the island fringes. Using the Sentinel 2 A multispectral data sets of the region, the coral reefs of the Bay of Bengal fringe shores and the Andaman Sea fringe coastal areas are estimated by spatial modelling techniques (such as Fuzzy membership for habitat modelling and Sen2Cor for coral reef habitats analysis) for the islands of the South Andaman group. The reef sedimentary landforms are well classified with the satellite data to identify their physiographic diversities, regional settings, and the status of the coral fringe islands. Fringing coral reefs are located on the east coast, and the barrier reef flats are located on the west coast of the island systems in the region. Similarly, coral reefs are also well developed and sheltered along the banks of the channels, straits and estuaries or ria type shores in the physiographic settings. Finally, the patch reefs are found into the lagoons, watery basins, tide pools and expanded rock pools of the shore platforms which are fringed with the South Andaman Islands.

The chapter also represents the function of the fringe corals as 'carbonate factory' in the production and supply of sediments to the shore platforms by considering the energy dimensions of the windward and leeward sides of the island groups. The shore platform morphodynamics are run by coastal cell circulation systems at the local scale and regional scale with a supply of carbonate sediments and their transportation paths from sources to sinks to produce the feedback mechanisms (both positive and negative) in the island fringe morphology. A few islands are well studied in this regard by monitoring geomorphology along the shore platform-based features and their morphometry.

Various karstification processes, such as limestone dissolutions, deposition of micrites from the saturated sea waters into the platforms, subaerial weathering systems, waves splashing, water layer weathering and salt weathering are identified through the morphological signatures and bio-encrustation zones in the carbonate platforms, and the previous denudational processes of the Island fringes to highlight the regional development diversity of karstic landforms. The role of sediment supply from the carbonate machines by the activities of tsunami waves, swell waves and wind waves, as well as the vertical displacement of the shoreline during seismo-tectonic events also played a role in island fringe progradation of the regional settings.

Antecedent Karst Morphology and Future Responses to the Changing Environment

Chapter 5 XE described the antecedent karst morphology in this section, in which the reef facies topographic patterns (highs and lows) of the islands are considered in stimulating the accumulation rates of reef sediments in paleo-oceanographic environments with sea level oscillations and the formation of reef margin benches at different depths. The antecedent platform theory also supported the mechanism of origin of coral reefs under past sea level variations in the tropical seas of the region. There are three different types of antecedent karst landforms in the region, such as, eroded reef terraces into the shore platforms with residual benches, exposure of limestone pavements with inherited karstic features visible after the removal of cover deposits by fluctuations of water levels at the shore fringe environment, and the identification of karstic features over the modern reef flats under shallow marine environment which were developed during the Last Glacial Maximum (LGM) after sub-aerial exposures around the oceanic environment. The presence of antecedent topographic features of island fringe environment hence influenced the growth and patterns of modern coral reef complexes in the region.

GEBCO bathymetry data is used to identify the antecedent topographic features of the island fringe environment in these sections. Various karst features of the island coasts are also explored to highlight their location as potential geo-tourism development sites in the region. The section also demonstrated how the coral fringe karstification of the islands responded to the physical impacts of the 2004 tsunami event and the great earthquake in Andaman.

Types of bio-encrustations and their ecological niches with topographic components are identified in the field survey methods, as well as their impacts are assessed on the carbonate platforms. Finally, the role and the nature of wooded islands, shore fringe platforms, reef banks, and reef flats are also considered to explain the regional settings and growth patterns of modern coral reef islands in the region.

References

Acharyya, A., & Dasgupta, S. (2006). Tectonic deformation due to 26 December 2004 earth quake-Revisited in Andaman. *Indian Mineralogical, 60*(3–4), 119–136.

Awasthi, N., Ray, J. S., Laskar, A. H., & Yadava, M. G. (2013). Chronology of major terrace forming events in the Andaman Islands during the last 40 kyr. *Journal of the Geological Society of India, 82*(1), 59–66.

Benumof, B. T., Storlazzi, C. D., Seymour, R. J., & Griggs, G. B. (2000). The relationship between incident wave energy and sea cliff erosion rates: San Diego County, California. *Journal of Coastal Research*, pp. 1162–1178.

Bland, W., & Rolls, D. (1998). *Weathering: An introduction to the scientific principles* (p. 288). Arnold Publishers.

Bird, Eric, C. F. 2008: Coastal geomorphology—An introduction 2nd Ed. John, Wiley and Sons Ltd.411

Bandopadhyay, P. C., & Carter, A. (2017). Geological framework of the Andaman-Nicobar Islands. *Geological Society, London, Memoirs, 47*(1), 75–93.

Bandopadhyay, P. C., & Ghosh, B. (2015). Provenance analysis of the Oligocene turbidites (Andaman Flysch), South Andaman Island: A geochemical approach. *Journal of Earth System Science, 124*(5), 1019–1037.

Berryman, K. R., Ota, Y., & Hull, A. G. (1989). Holocene paleoseismicty in the fold and thrust belt of the Hikurangi subduction zone, eastern North Island, New Zealand. *Tectonophysics, 163*, 185–195.

Bögli, A. (2012). Karst hydrology and physical speleology. Springer Science & Business Media

Curray, J. R. (2005). Tectonics and history of the Andaman Sea region. *Journal of Asian Earth Sciences, 25*(1), 187–232.

Chatterjee, P.K. (1967): Geology of the main Islands of the Andaman area, In *Proceeding Symposium on Upper Mantale Project*, N.G.R.I, Hyderabad, India, Session V, pp. 348–362

Das, D. P., & Pramanik, K. (2011). Evaluation of impact of earthquake and tsunami on the coastal morphology of Andaman-Nicobar Island using multi-sensor temporal satellite data: Indian jour. *Indian Journal of Geosciences, 65*, 9–22.

Daneshian, J., Sarkar, S., & Sharma, V. (2007). Miocene Palynoflora from Inglis Island, Andaman Sea and its Palaeoecological Implication. *Journal of Geological Society of India (Online archive from Vol 1 to Vol 78), 70*(1), 147–157

Ellison, J. C. (1994). Caves and speleogenesis of Mangaia, Cook Islands. Atoll Research Bulletin

A. Paul et al., *Karstified Coasts of the South Andaman Islands*,
https://doi.org/10.1007/978-3-032-10299-7

Ford, D., & Williams, P. D. (2013). Karst hydrology and geomorphology. Jhon Wiley and Sons

Krishnan, M. S. (1982). Geology of India and Burma CBS Publishers. 1967

Karunakaran, C., Pawde, M. B., Raina, V. K., Ray, K. K., & Saha, S. S. (1964). Geology of the south Andaman Island, India. In *Reports 22nd International Geological Congress*, New Delhi, India, XI, vol. 79, p. 100

Kishore, S., Jauhri, A. K., Singh, S. K., Malakar, B., & Misra, P. K. (2015). Coralline algae from the Neill West Coast Formation (Pleistocene), Neil Island, South Andaman, India. *Journal of the Palaeontological Society of India, 60*, 57–69.

Lace, M. J., & Mylroie, J. E. (2013). Coastal karst landforms, CRL-5, Springer, p. 425

Lajoie, K. R. (1986). Coastal tectonics. *Active Tectonics*, pp. 95–124

Mylroie, J., Mylroie, J., Humphreys, W., Brooks, D., & Middleton, G. (2017). Flank margin cave development and tectonic uplift, Cape Range, Australia. *Journal of Cave and Karst Studies, 79*(1)

Ota, Y., & Yamaguchi, M. (2004). Holocene coastal uplift in the western Pacific Rim in the context of Quaternary uplift. *Quaternary International, 120*, 105–117.

Paul, A. K. (2005). *Tsunami: An assessment of the Disaster over the nations in the Indian Ocean* (pp. 1–125). India ACB Publications, Kolkata.

Paul, Ashis, K. R. (1998). Problems of groundwater contamination and water scarcity in Andaman and Nicobar Islands DELF, New Delhi, pp. 13–33

Palmer, A. N. (1991). Origin and morphology of limestone caves. *Geological Society of America Bulletin, 103*(1), 1–21.

Palmer, A. N. (2007). Cave geology and speleogenesis over the past 65 years: Role of the National Speleological Society in advancing the science. *Journal of Cave and Karst Studies, 69*(1), 3–12.

Peltier's weathering model. (1950). https://www2.geog.soton.ac.uk

Paul, A., Bandyopadhyay, J., & Paul, A. K. (2018a). Geomorphological mapping and environmental zoning approach to coastal management in Havelock Island, South Andaman, India. In *39th Asian Conference on Remote Sensing: Remote Sensing Enabling Prosperity, ACRS*, pp. 346–357

Paul, A. K., Kamila, A., & Ray, R. (2018). Natural threats and impacts to mangroves within the coastal fringing forests of India. *Threats to Mangrove Forests: Hazards, vulnerability, and management*, pp. 105–140

Paul, A., Bandyopadhyay, J., & Paul, A. K. (2017). Monitoring vertical cliffs, embayments and shore platform morphology in parts of Neil Island, South Andaman using spatial information technology. In *38th Asian Conference on Remote Sensing–Space Applications: Touching Human Lives, ACRS 2017*

Paul, J., Rajendran, C. P., Lowry, A. R., Andrade, V., & Rajendran, K. (2012). Andaman postseismic deformation observations: Still slipping after all these years? *Bulletin of the Seismological Society of America, 102*(1), 343–351.

Paul, A., & Paul, A. K. (2023a). Geomorphological Analysis of the coral fringed coasts of Andaman and Nicobar Islands using geospatial techniques. In *Crisis on the Coast and Hinterland: Assessing India's East Coast with Geomorphological, Environmental and Remote Sensing and GIS Approaches*, pp. 303–325. Cham: Springer Nature Switzerland

Paul, A. K., & Paul, A. (2023b). Geomorphological diversity and sea level rise vulnerabilities on India's East Coast. In *Crisis on the Coast and Hinterland: Assessing India's East Coast with Geomorphological, Environmental and Remote Sensing and GIS Approaches*, pp. 3–25. Cham: Springer Nature Switzerland

Pillai, C. S. G. (1983). Coral reefs and their environs. *CMFRI Bulletin, 34*, 36–43.

Rajendran, K., Rajendran, C. P., Earnest, A., Prasad, G. R., Dutta, K., Ray, D. K., & Anu, R. (2008). Age estimates of coastal terraces in the Andaman and Nicobar Islands and their tectonic implications. *Tectonophysics, 455*(1–4), 53–60.

Rajendran, C. P., & Rajendran, K. (2022). Earthquakes of the Indian Subcontinent. Springer Singapore

Rabinowitz, D. (1978). Dispersal properties of mangrove propagules. *Biotropica*, pp. 47–57

Roy, S. D., George, G., Soundararajan, R., Krishnan, P., Murugesan, S., & Kaliyamoorthy, M. (2014). Status of coral reefs in Andaman after two major catastrophic events—Tsunami of December 2004 and bleaching of May 2005. *Ecology Environment and Conservation, 20*, 539–544.

Ray, K. K. (1982). A review of the geology of Andaman and Nicobar islands. *Miscellaneous publication-Geological survey of India*, pp. 110–125

Rodet, J., Meyer, R., Dupont, J. P., Sayaret, D., & Tomat, A. (1995). Relations entre la dissolution des carbonates et le remplissage terrigène dans le karst de la craie en Normandie (France). Comptes rendus de l' Académie des sciences. Série 2. *Sciences de la terre et des planètes, 321*(12), 1155–1162

Strahler, A. N. (1960). *Physical geography*. Wiley.

Sharma, V., & Srinivasan, M. S. (2007). *Geology of Andaman-Nicobar: The Neogene* (p. 163). Capital Publishing Company.

Singh, P., & Vimal, K. P. (1976). Origin and geology of Neill Island. South Andaman India. *Journal Palaeontological Society India, 19*, 28–36.

Singh, P. R. A. T. A. P., Vimal, K. P., & Nautiyal, D. D. (1978). Early Pliocene Diatoms and Silicoflagellates from Neill Island, South Andaman, India, PART-I. *Journal of the Palaeontological Society of India, 21*, 51–61.

Srinivasan, M. S., & Azmi, R. (1976a). Contribution to the stratigraphy of Neil Island, Ritchie's Archipelago, Andaman Sea Proceeding VI, Indian Colloq. Micropal. And strat. pp. 283–301

Srinivasan, M. S., & Chatterjee, B. K. (1981). Stratigraphy and depositional environments of Neogene limestone of Andaman Nicobar Islands; Northern Indian Ocean. *Journal Geological Society India, 22*, 536–546.

Sharma, J. S., & Kar. A. (2013). Ground water information Booklet, South Andaman District Andaman & Nicobar Islands CGWB, pp. 1–42

Sneed, E. D., & Folk, R. L. (1958). Pebbles in the lower Colorado River, Texas a study in particle morphogenesis. *The Journal of Geology, 66*(2), 114–150.

Trudgill, S. T. (1987). Bioerosion of intertidal limestone, Co. Clare, Eire—3: Zonation, process and form. *Marine Geology, 74*(1–2), 111–121

Vohra, C. P., Haldar, D., Ghosh Roy, A. K., & Ghose, N. C. (1989). The Andaman–Nicobar ophiolite complex and associated mineral resources–current appraisal. Phanerozoic ophiolites of India, pp. 281–315

Vita-Finzi, C., Hidayat, S., 1991. Holocene uplift in West Timor. *Journal Southeast Asian Earth Science*, pp. 387–393

Viles, H. A. (Ed.). (1988). Biogeomorphology. B. Blackwell

Index

A. Paul et al., *Karstified Coasts of the South Andaman Islands*, https://doi.org/10.1007/978-3-032-10299-7

Index

A. Paul et al., *Karstified Coasts of the South Andaman Islands*, https://doi.org/10.1007/978-3-032-10299-7

MIX
Papier aus verantwortungsvollen Quellen
Paper from responsible sources
FSC® C105338

If you have any concerns about our products,
you can contact us on
ProductSafety@springernature.com

In case Publisher is established outside the EU,
the EU authorized representative is:
Springer Nature Customer Service Center GmbH
Europaplatz 3, 69115 Heidelberg, Germany

Printed by Libri Plureos GmbH
in Hamburg, Germany